Spillways on River Levees

Gérard Degoutte and Rémy Tourment
Editors

Éditions Quæ

This book was translated by Diana Huet de Guerville from the French edition entitled
Les déversoirs sur digues fluviales.

To cite the book: G. Degoutte, R. Tourment (eds), 2021. *Spillways on River Levees.*
Versailles, Ed. Quæ, 184 p. DOI: 10.35690/978-2-7592-3285-7

Éditions Quæ
RD 10, 78026 Versailles Cedex (France)
www.quae.com - www.quae-open.com

© Éditions Quæ, 2021
ISBN (print): 978-2-7592-3284-0
ISBN (PDF): 978-2-7592-3285-7
ISBN (ePub): 978-2-7592-3286-4

Contents

Foreword

This handbook was produced at the initiative of the French Ministry in charge of the environment and written by a working group led by INRAE and supervised by the DGPR/SRNH/STEEGBH (General Directorate for Risk Prevention/Natural and Hydraulic Risks Department/Technical Department for Electrical Energy, Large Dams, and Hydraulics).

The handbook was first published in French in 2012 and initially translated into English the same year, but not published. Gérard Degoutte, Paul Royet (INRAE, retired), and Aline Leclerc (Agora Traductions) worked on the first version of the translation. In 2021, Rémy Tourment (INRAE, chairman of the ICOLD Technical Committee on Levees) updated the English version of the handbook with the help of Adrien Rulliere (INRAE) and translator Diana Huet de Guerville.

The working group in charge of the French version, initially led by Rémy Tourment (INRAE) and then by Gérard Degoutte (CGAAER and INRAE), also included the following members:

- Nicolas-Gérard Camphuis (CEPRI then the Loire-Brittany Water Agency), general reviewer
- Gérard Degoutte (CGAAER and INRAE), group leader and general reviewer
- Patrick Deronzier (MEDD/DEEEE), during the first meetings
- François Dols (Director of the Rhône-Méditerranée Water Agency) during the first meetings
- David Goutx (Agro ParisTech then Météo France), co-writer of Chapter 4
- Frédéric Grelot (INRAE), during the first meetings and the review of Chapter 8
- Xavier Martin (CGEDD), during the first meetings and reviewer of an initial version
- Nicolas Monié (MEDDTL/DGPR/SNRH/STEEGBH), co-writer of Chapter 2
- Jean Maurin (DREAL Centre), co-writer of Chapters 3 and 7, writer of Appendix 1, and general reviewer
- Paul Royet (INRAE), main writer of Chapter 6 and general reviewer
- Rémy Tourment, group leader, general reviewer, and coordinator of the second English version
- Gilles Trapatel (CNR), c-writer of Appendix 2
- Yves Le Trionnaire (MEDD/DE), during the first meetings
- Thibaut Mallet (Symadrem), co-writer of Chapters 4 and 7

The following people also helped write the French version of this handbook:
- Francis Fruchart (CNR), Chapter 4 and Appendix 2
- Frederic Laugier (EDF) for the PK weir paragraph update in Chapter 5
- Patrice Mériaux (INRAE) and Yann Quefféléan (ONF/Direction technique RTM), for the torrential river paragraph in Chapter 5
- Stéphane Bonelli (INRAE) and Daniel Puiatti (DPST consulting), for the lime treatment paragraph in Chapter 5

We thank the following people for their proofreading:
- Jean-Marc Kahan (MEDDTL/DGPR/SNRH, Director of STEEGBH) and Patrick Ledoux (CETE Méditerranée), for the entire handbook
- Christophe Peteuil (CNR), for Chapter 5

We would like to thank the following people for their illustrations or the information they provided:
- Stéphane Barelli (VNF Nancy), Arnaud Bollery (CU Montbéliard), Pierre Cadoret (DDTM 11), Benoît Cortier (Hydratec), Rémy Croix (Egis Eau), Georges Darbre (Swiss Committee on Dams), David Di Dio Basalmo (DDT 67), Francis Fruchart (CNR), Jérôme Gondran (SMAV Durance), Olivier Manin (Symbhi), Olivier Overney (Ofev Bern), Franck Rangognio (BRL Ingénierie), Burkhard Rosier (EPFL), Akim Salmi (ISL), Xavier Suisse de Sainte-Claire (Egis Eau), and Éric Vuillermet (BRL Ingénierie).

This handbook was funded as part of the RESBA project, financed by the Interreg V-A France-Italy (ALCOTRA) Cooperation Programme.

Le guide a été financé dans le cadre du projet RESBA, financé par le programme de coopération INTERREG V A France-Italie ALCOTRA.

Introduction

The likelihood that overflow will occur on flood protection levees is far from insignificant. When water overflows an earthen levee, it greatly increases in velocity and erodes the levee slope or toe. This erosion extends backwards and opens up a breach that provokes sudden flooding in the supposedly protected area, often with significant economic impact. The suddenness of this process can lead to human casualties in areas with inhabitants or transport infrastructure.

Levees are a double-edged sword: they offer protection against medium-sized floods if properly built, but can create a hazard during high floods if no provisions are made to secure them against overtopping.

Dams almost all have flood spillways to prevent even extremely rare floods from overflowing over their crests. **Can we systematically apply this technique to levees, particularly earthen ones?**

This technical handbook outlines the benefits and limitations of spillways on flood protection levees. Some spillways can protect leveed areas by limiting the adverse consequences of overflows. They are meant to be used infrequently. Other spillways are specifically designed for flood control and are used on a more regular basis. Since these two types of spillways have very different objectives, we will refer to the former as safety spillways and the latter as diversion spillways. Some structures, especially the oldest, actually play both roles. The hydraulics remain the same regardless, and this handbook covers all types of levee spillways.

What types of levees? This handbook examines flood protection levees on rivers, including mountain torrents. We will not address levees along coasts or canals. We focus on existing levees that already have spillways or the possibility of adding them, as well as new levee projects that could incorporate spillways into their construction design.

1

Background, definitions, various configurations

Gérard Degoutte

This chapter provides an overview of river levees, levee systems, and their potential spillways. We will first describe the various configurations before addressing their purpose. A spillway's function is closely tied to the purpose of the area it protects or fills with water, with a broad range of scenarios we will examine in closer detail.

Typology of flood protection levees on rivers

Definition

Levees are typically very long structures built above the natural ground level to either channel water or prevent it from flowing through. Excluding those built along canals, levees are designed to protect certain areas against river or marine flooding. This handbook only addresses riverine flood protection levees that provide partial or total protection against floodplain inundation.

As we will see later, levees may protect against certain floods while only delaying stronger ones. But this is a consequence of the definition rather than part of the definition. The main issue is that the presence of a river levee is a double-edged sword: it offers protection against small- or medium-sized floods but creates a hazard during strong or extreme floods or when it is poorly maintained. If the levee fails, the resulting flood wave may cause even more damage in the floodplain than if there had been no levee at all. Earthen levees offer particularly low resistance to overflows. They may also breach before overflowing because of internal erosion, toe scour, or sliding.

These levees (sometimes called "dry levees") are seldom subject to hydraulic loading, much like flood retention dams or basins.

River levees are generally made from earthfill taken from the riverbed or the surrounding area. They are sometimes built of masonry or concrete, particularly on urban sites, in which case they are sometimes called dikes.

Earthen dams are also sometimes called levees, but this term is incorrect and should be avoided. A dam is built across a river to block at least the riverbed, and frequently the floodplain and beyond. A levee, on the other hand, never blocks the riverbed.

Levees may connect several natural topographical features such as hillsides, promontories, or terraces. Although these natural features are not considered levees, we should analyse their resistance in the same way as human-made levees. We must clearly define the area that is being protected from flooding by one or more levees and potentially some natural features. All these components taken together form what we call a leveed system.

Longitudinal levees

In leveed valleys, the goal has often been to prevent water from entering most of the floodplain. The levee runs parallel to the watercourse, either at the riverbank level or some distance away. In the latter case, the space between the levee and the riverbank is called an unprotected floodplain or *ségonnal* in French (Figure 1.1). As its name suggests, this area is not protected by the levee and will be inundated by even higher water levels if flooding occurs.

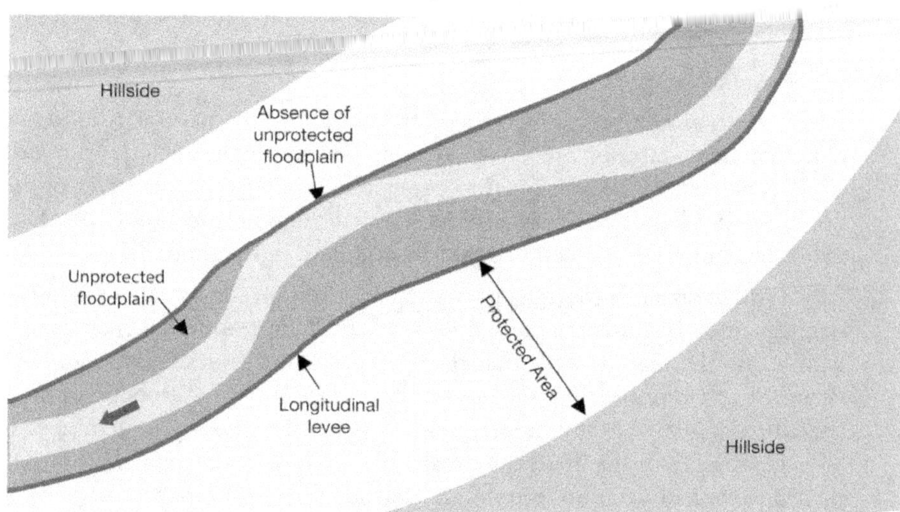

Figure 1.1. Longitudinal levee and unprotected floodplain.

Close protection levees

Levees that prevent large areas from being submerged will worsen overtopping downstream and on the opposite bank. This is why close protection levees are built as near as possible to the stakes to be protected, thereby expanding the flooded area (Figure 1.2). If the stakes to be protected are next to the hillside, the close protection levee will close onto it. If all the stakes to be protected are in a flood-prone area, the close protection levee may be called a ring levee.

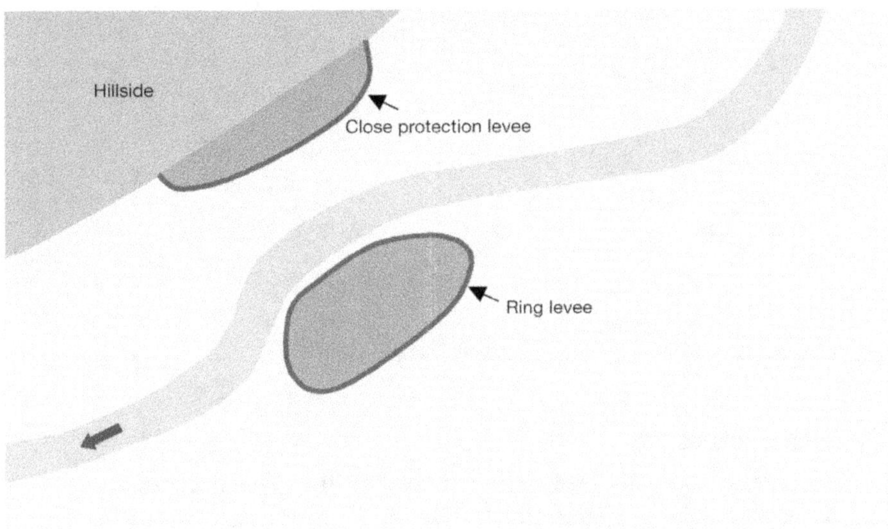

Figure 1.2. Close protection levees completely or partially surround areas to be protected.

Levee rank

A close protection levee may complement a longitudinal levee: a primary levee protects mainly rural areas and a secondary close protection levee protects mainly urban areas. The rank of protection refers to the order in which levees go into effect in the event of a strong flood. No other consideration, either in terms of volume or height should be inferred. In Figure 1.3, on the right bank, the "secondary levee" designation is clear. The area on the left bank is protected from overflows by the longitudinal levee. A peripheral levee prevents discharge into the floodplain from a tributary, a spillway, or an upstream breach. In this configuration, it is hard to say which levee will be activated before the other, so no particular rank of protection is given.

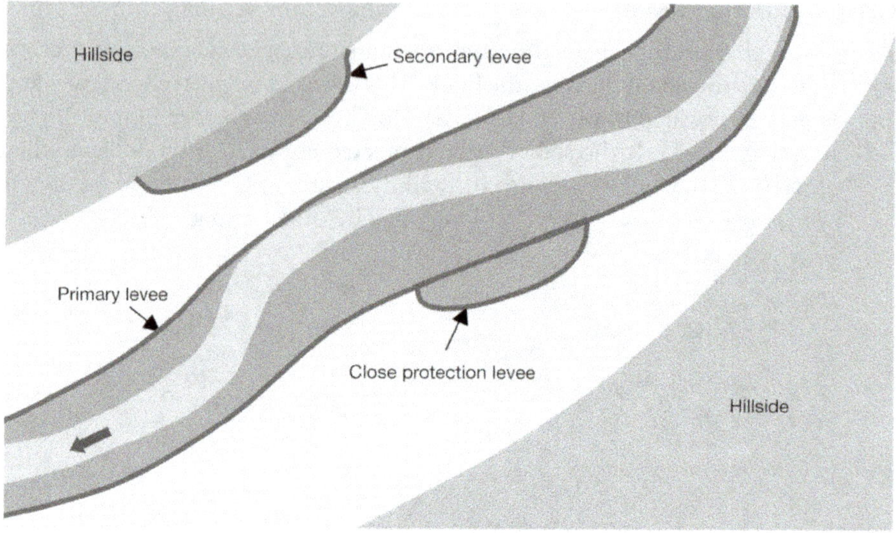

Figure 1.3. Primary and secondary levees where applicable.

Transverse levees

Transverse levees are built perpendicular to the general direction of the valley. They complement longitudinal levees and protect against upstream or downstream inflows. In Figure 1.4, on the left bank, two transverse levees close onto the hillside and isolate a protected area. The upstream levee prevents water from flowing into the flood plain and the downstream levee prevents water from flowing from downstream due to the backwater effect. On the right bank, the transverse levee does not enclose a protected area but splits it into two parts. This may help protect the downstream section if there is a breach upstream. This type of levee may also separate two leveed areas with different purposes: a flood expansion area upstream and a flood protection area downstream. One example is the La Montagnette levee between Vallabrègues and Tarascon, shown in Figure 1.14, or the canal levee between Tours and Saint-Pierre-des-Corps.

Protected areas and flood expansion areas

We define two types of leveed areas in very different contexts: protected areas in urban environments and flood expansion areas that are preferably non-urbanised. As we will see later, an area can be either a protected area (up to a certain water level) or a flood expansion area (above this level). We use the term "leveed system" to mean either of these. A leveed system is an entire area in a floodplain that has some level of flood protection thanks to one or more levees or natural topography (such as the valley slope).

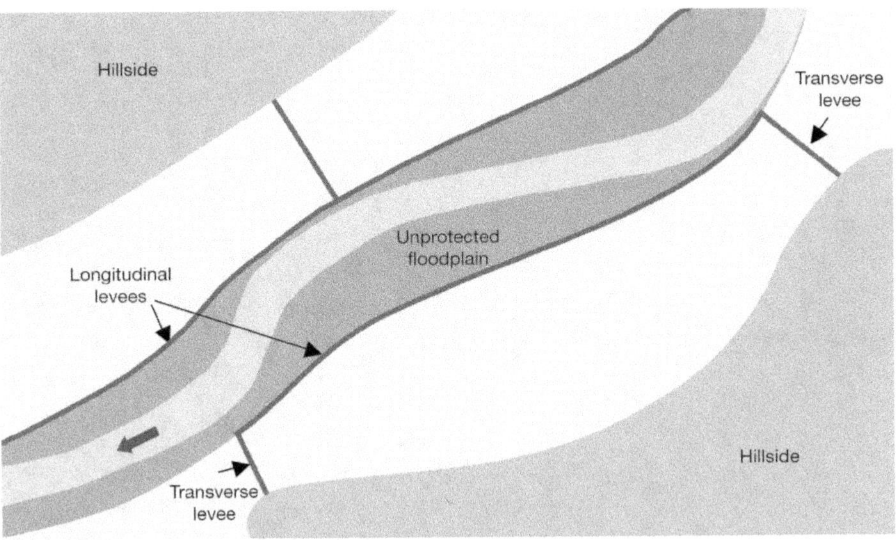

Figure 1.4. Transverse levees.

Protected area

The term "protected area" is quite standard in France. It was used in French Decree 2007-1735 of 11 December 2007 on the safety of hydraulic works, without being properly defined. The *Environnement* circular of 8 July 2008 on this decree includes the following definition:

"A protected area is the area that will not be inundated by a flood that reaches the structure's protection level. It is not the more limited area in which residents would be endangered by high or fast water levels in the event of a levee failure. Neither is it an area that is inundated for the design flood level in the PPRI (Flooding Risk Prevention Plan) by known high water levels, a 100-year flood, or the maximum floodable area."

The more recent French Decree 2015-526 of 12 May 2015 also mentions protected areas; a protected area is associated with a levee system and a protection level. A protected area can also include specific parts associated with different protection levels.

A protected area is a contiguous surface that is protected from flooding by a set of levees or other structures (road embankments, etc.) or a raised topographical feature such as a hillside or a terrace. It is an area that is likely to flood without a levee and is protected from flooding as long as the levee plays its protective role – that is until the levee is overtopped or damaged. See Figure 1.5. For close protection levees, as shown in Figures 1.2 and 1.3, the protected area is clearly defined.

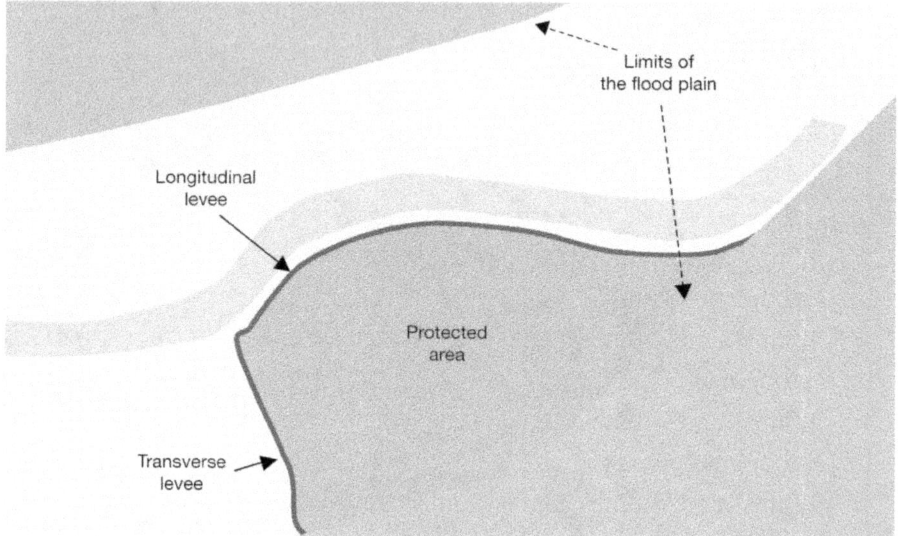

Figure 1.5. Area protected by a longitudinal levee, a hillside, and a transverse levee. This configuration will be effective whatever the direction of the river current.

In Figure 1.5, the protected area is closed and must include a drainage system.

Another option is for the protected area to be partially open downstream. See the example in Figure 1.6. This area is closed by the levee on one side and cut off by a tributary flowing on the other side through the hillside. This will allow common floods to fill in the area from downstream through the backwater effect. The entire area is not a protected area since a small section at the junction is flood-prone, while the rest is protected by the levee.

Most of the protected areas that have been built recently or are currently planned are in highly urbanised areas. After all, there is no need to protect an area in which flooding will have limited impact, or in which there are no stakes. In the past, however, mainly rural areas were protected. Camargue Island is probably the best example, along with the Isère River Valley upstream from Grenoble, the Save River Valley in Gers, the Vidourle River Valley, and many more.

Flood expansion area

Our definition of a flood expansion area comes from the French water portal (www.eaufrance.fr) and French Master Plan for Water Development and Management: "A flood expansion area is a natural or leveed area into which water spreads when a watercourse overflows into the floodplain. Temporary water storage attenuates the flood by extending its flow time. This storage helps aquatic and land ecosystems function properly. A flood expansion area typically refers to areas with no or limited urbanisation and development."

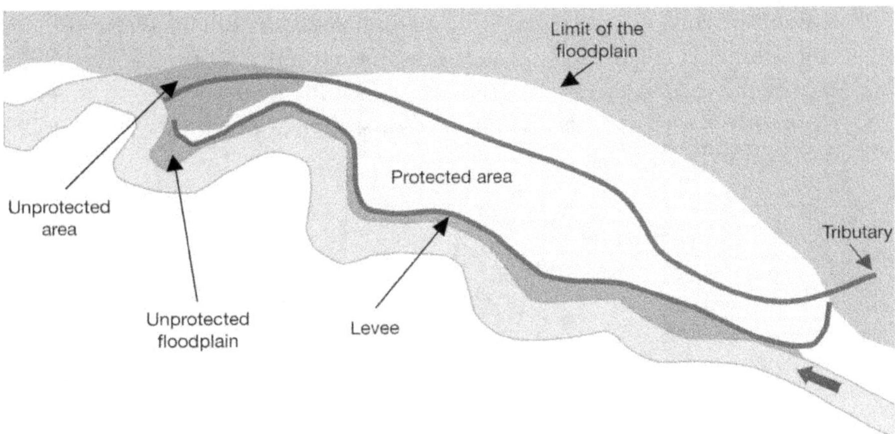

Figure 1.6. Open protected area: a downstream tributary flows through the levee.

Even though it is not the focus of this handbook, this definition shows that natural (therefore non-leveed) riverine floodplains are flood expansion areas. They are not constrained by structures such as levees. The construction of longitudinal levees is designed to limit flood expansion, at least for all floods that cannot flow over the levee crest or destroy it. Efforts to add levees or backfill to the floodplain limit the expansion areas and increase peak discharge downstream, requiring more and more construction features.

This is why priority should be given to flood retention structures or measures. In this context, the areas designed to be flooded are also flood expansion areas. To differentiate between these two cases, we will use the terms natural vs. controlled flood expansion areas to avoid any confusion. Other acceptable, though more infrequent terms are also used for controlled flood expansion areas such as controlled flood fields in the Isère River Valley and dynamic flood retention areas.[1]

Natural and controlled flood expansion areas can both contribute to a dynamic flood retention strategy, as do other structures such as flood attenuating dams, diversion basins, setting back or removing levees, floodplain forestation, installing hedges perpendicular to the direction of the flow in the floodplain or catchment area, and any other changes to the area. In the rest of the handbook, we will use the term (natural or controlled) flood expansion area rather than dynamic retention for accuracy, even though flood expansion areas do perform dynamic retention.

Flood expansion may have a hydraulic or ecological function or both. The goal is to reduce downstream flooding through temporary water storage. This also preserves the ecological value and diversity of the floodplain.

---

1. For more information on dynamic retention, see the publication by the French Environment Ministry (Chastan et al., 2004).

A flood expansion area is based on natural contours but may also use artificial embankments (Figure 1.7). Water flows in naturally when the river is not leveed. If the river is leveed, water flows artificially over a weir on the levee. Less frequently, water may also be pumped or siphoned in (in this case, a spillway is not required).

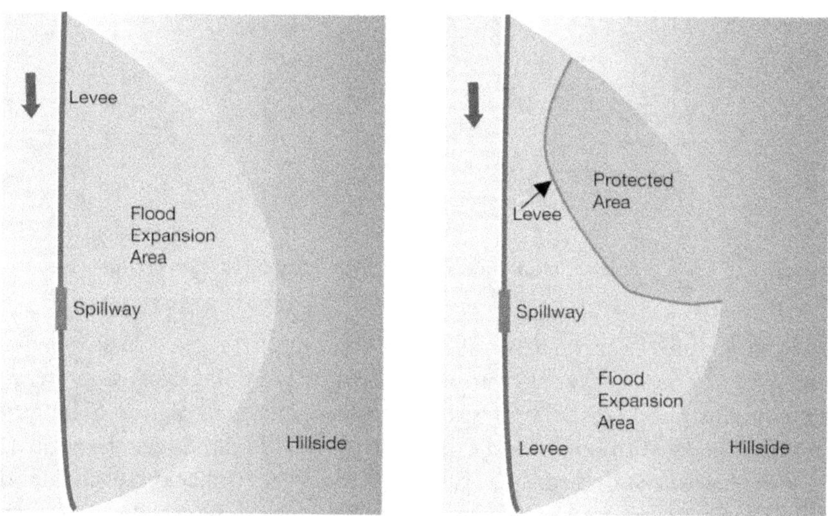

Figure 1.7. Controlled flood expansion area fully enclosed by natural flood expansion features on the left; limited by a protected area on the right.

Figure 1.14 below shows an example from the Vallabrègues flood expansion area on the Rhône River.

In a flood expansion area based on natural features, no overflow occurs compared to an area without a levee or spillway (Figure 1.7, left). However, in a flood expansion area whose floodable footprint has been strongly reduced by borrow materials or levees, overflows might occur (Figure 1.7, right).

Controlled flood expansion areas are generally non-urbanised (or only slightly urbanised). They can be agricultural, forest, or ecological areas where some sports or leisure activities may also be possible.

Maintained or reactivated flood expansion areas help temper downstream floods (making them weaker and more spread out). The weaker the downstream flow, the lower the waterline. When the flow is subcritical, waterlines are also lower over a certain distance upstream, via the backwater effect. We will return to these hydraulic aspects in Chapter 3.

A controlled flood expansion area is also a protected area. As long as the spillway does not overflow into the flood expansion area, this area is protected by the levee. Of course, a natural flood expansion area is not a protected area (since flooding occurs there).

Characteristics of flood expansion areas within protected areas

Here we are only referring to controlled flood expansion areas, meaning those with levees. These are also protected areas, but with a specific purpose.

For now, we will ignore what these areas are called and focus on the developer's goal, such as optimal flood attenuation. Developers may want the area to be flooded relatively early, as long as there are limited stakes. If necessary, isolated residential areas can be offered close protection. This is typical in a leveed flood expansion area (which is also a protected area).

Developers may want to protect a densely populated part of the flood expansion area, using a close protection levee to minimise the reduction of the natural flooding area and limit both upstream and downstream impacts. This is typical of a protected area in the strictest sense. Protection should be as close as possible to the stakes. The purpose is not to attenuate floods, even though a slight attenuation could be useful.

Developers may also have a dual purpose: flood attenuation and protection of an area with important stakes. This is surely what inspired the Comoy levee project two centuries ago (see Appendix 2, "A history of spillways on the Loire River"). Before that, most of the Loire Valley was composed of protected areas with extensive rural spaces. Flood attenuation was not the objective, on the contrary, since at the time it was thought that levees could be non-submersible. This is why some large protected areas do not have close protection levees. For instance, the Authion leveed area extends over 200 km^2 and now protects 50,000 inhabitants. It does not have a spillway. The leveed areas in which Comoy spillways were built are still protected areas, but their impact on flood attenuation is significant, especially through cumulative effects. They can therefore be considered protected areas as well as flood expansion areas. This is the case of the Ouzouer leveed area, with 5,000 inhabitants in 65 km^2.

It is important to remember that the objectives were very different in the first two cases (optimal flood attenuation and protection of a densely populated area). In a protected area with high stakes, any inflow of water should be avoided, whether from the river overflowing, levee overtopping, or levee failure. Most often, spillways in these areas only start operating above a 100-year flood level. Conversely, in controlled flood expansion areas, overflows are encouraged at the ideal time to ensure suitable flood attenuation. Typically, flood expansion areas are flooded for 10- to 50-year floods, meaning far more frequently than in protected areas.

Despite their differing purposes, these areas are all similar in terms of overflow. They follow the same hydraulic rules:
– A controlled flood expansion area first serves as a protected area for moderate floods that do not reach the spillway crest.
– It then becomes a flood expansion area and helps protect downstream areas; a flood expansion area is designed to protect both downstream areas and its own territory.

– Once a high-stakes protected area is flooded by the spillway, it necessarily becomes a flood expansion area; if it covers a large surface, it can help attenuate flooding. This is also the case for a protected area without a spillway if there is a breach in the levee.
– Lastly, some areas may be considered both flood expansion areas and protected areas, such as the Loire leveed areas mentioned above. The Camargue is currently a protected area, but it could also be used for flood expansion in the future.

However, these areas are very different in terms of spatial planning. A protected area without a flood attenuation objective is most likely urbanised, whereas a flood expansion area is generally non-urbanised.

Making a case-by-case distinction seems critical for spillway design. In Chapter 3, we will look separately at the design of diversion spillways in flood expansion areas and safety spillways in protected areas. When an area is both a protected area and a flood expansion area, designers face additional constraints since it is impossible to optimise both protection and flood attenuation.

Of course, nothing prevents us from reasoning at the watershed level:
– A flood expansion area on its own is not that beneficial. Only an entire set of natural or controlled flood expansion areas will have a significant impact on flooding.
– A protected area is a unit of protection that can be effective on its own, but a set of protected areas could have a negative impact by reducing the floodable area too strongly.

In summary

The floodplain of a leveed valley may include:
• Protected areas designed to reduce flooding frequency and, if possible, protect the area against rare events that may cause overflows. To keep from reducing the floodable area too sharply, the protected area should not extend too far – meaning the protection should be placed close to the stakes.
• Natural flood expansion areas, where flooding is permitted, to prevent any increase in downstream flood flows; we will not cover this topic.
• Controlled flood expansion areas where flooding is facilitated to reduce flood discharge downstream and/or for ecological purposes.
• Protected areas where flood attenuation is also intended, and which may be referred to as a protected area or flood expansion area, depending on the topic.

This handbook is primarily focused on spillways in protected areas in the strictest sense, as well as spillways in controlled flood expansion areas.

The fragility of levee systems

Earthen levees may experience damage from a variety of causes, mainly internal erosion, external erosion in the case of overflows, scour with levees at riverbank level, or collapse of the slope. These mechanisms can lead directly to a breach, or create a chain reaction in the form of processes (scenarios) that can also lead to a breach. A well-designed and well-built earthen levee that is also properly monitored and maintained should be able to resist these effects, except overflow erosion.

Unlike dams, the probability that a levee will overflow is significant, at least in France. This typically occurs about 10^{-2} per year or more. As soon as levee overflow begins, water flows very quickly over the landside slope and erodes it by wearing particles away and carrying them down. The shear stress applied by a 1 cm water layer on a 1:3 slope is:

$$\tau_0 = \gamma_w.h.i = 9{,}810 \times 0.01 \times 0.316 = 31 \text{ Pa.}$$

Yet the acceptable shear stress on the ground ranges from 3 to 30 Pa. Therefore, even a low level of discharge can wear particles away. This eliminates the stabilising force they provide to upstream particles, which can, in turn, be worn away. This mechanism is regressive.

In the case of compact and non-cohesive soil, erosion begins at the base of the slope, where the water reaches maximum speed and abruptly changes direction. The jet effect erodes the soil at the base of the structure. When the flow has continued long enough, the water level rises downstream, resulting in a hydraulic jump (see green dotted line 1 in Figure 1.8a). The downstream water cushion will reduce the water's energy on impact, but not enough to prevent erosion. This depression increases by capturing flow from both sides, gradually creating a real channel that moves upward due to backward erosion, as shown in Figure 1.8a (from green to red and then purple dotted lines).

If this overflow continues, the crest width will gradually diminish (dashed line). If the flood does not last long, the process ends here: the entire width of the crest is not gashed, as shown in Photo 1.1. No breach occurs and there is no disastrous impact downstream. However, if the flow continues, the channel will be in a critical condition (see purple dotted line 2 in Figure 1.8a) in which the crest is completely eroded. The river surges into the channel, provoking a breach that marks the beginning of the levee's failure. From this point on, water from the river continues to erode the levee, even if the flood has receded. It quickly erodes the walls of the breach, which expands to the natural ground level at the levee toe (Photos 1.1 to 1.3). It generally creates a large erosion pit. When the breach reaches the levee toe, it then extends widthwise until a balance is reached: the upstream level drops and the downstream level rises and stabilises the breach. Nevertheless, the water will keep flowing through the breach unrelentingly until it returns to the riverbed. At the end of the process, the sides of the breach are relatively vertical; this fragile equilibrium is due to suction from the interstitial waters between the particles.

For loose and non-cohesive soil, erosion can start as soon as the water gathers speed, meaning at the downstream end of the crest (Figure 1.8b). The small cavity (see green dotted line 1 in Figure 1.8b) that is created will capture flows from the sides and extend downstream, forming a gully (see red dotted line in figure 1.8b). The gully also extends backwards, i.e., upstream, and flows towards the levee crest. If the flood event lasts long enough, the channel will extend past the downstream corner of the crest and create a breach, as in the previous example see purple dotted line 2 in Figure 1.8b.

With cohesive soil, erosion also starts at the toe with an erosion pit created by the jet effect, where the flow changes direction (with or without a hydraulic jump). The pit broadens along the sides and upstream. A channel is formed backwards with a staircase (or stepped) profile in a process called headcutting[2] (Figure 1.8c). Water will cascade down: horizontal jets dig erosion pits dissipate the energy of the water. Blocks of cohesive soil that form the edge of a step create an overhang through jet erosion (Figure 1.8d); they are knocked over and carried away by the flow before disintegrating. Through this process, the steps retreat upstream, increasing in height, and decreasing in number. When the upstream corner of the crest is reached (see purple dotted line 2 in Figure 1.8c), a breach will occur as in the previous examples.

In all three instances, the breach almost always reaches the levee toe and discharges a large volume of water into the plain.

Photos 1.1 and 1.2. Overtopping of two small earthen dams of similar construction; of short duration on the left and long duration on the right. On the right, there are stepped shapes. Though they are not levees, the mechanism is the same. (Photos: Paul Royet, 1983)

Photo 1.3. Breach on the left bank of the Vidourle River in Aimargues on 9 September 2002 (4 m × 25 m, 294 m³/s as indicated by SAFEGE modelling for SIAV). In the foreground, the top of a "step" has been cut out. (Photo: Aimargues town hall)

---

2. Hanson G. J., Robinson K. M., Cook K. R., 2001-Prediction of headcut migration using a deterministic approach, transactions of the ASAE, vol. 44(3), 525-531.

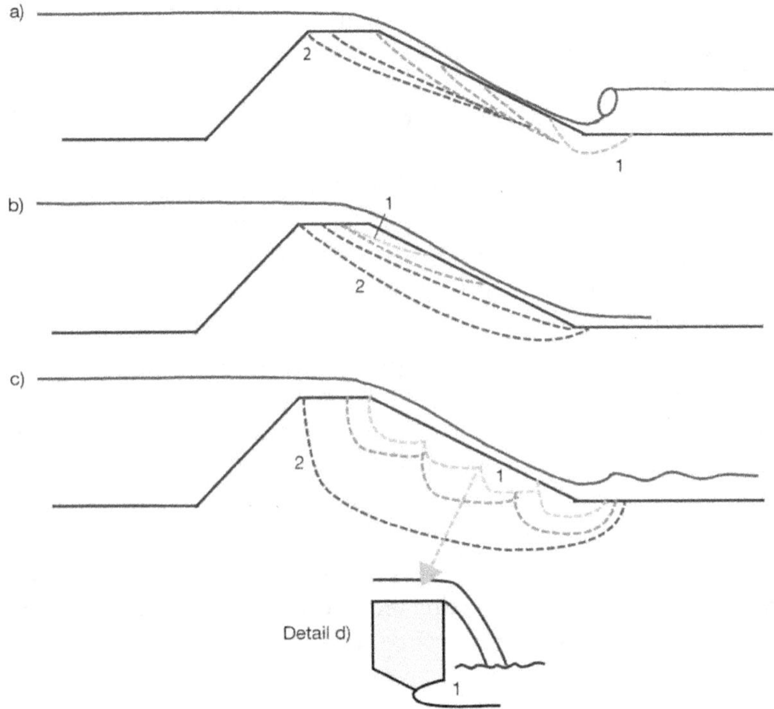

Figure 1.8. Principle of erosion via overflow of an earthen levee; (a): compact powdery soil; (b): loose powdery soil; (c) and (d): cohesive soil. 1) beginning of the erosive process; 2) beginning of the levee breach.

Overflow of an earthen levee most frequently leads to erosion that beings at the levee toe. However, erosion may sometimes occur on the landside slope after the crest overflows. Flow rate or shear stress is most important in determining when and where erosion begins. But the beginning of erosion does not necessarily lead to a breach. For a breach to form, flooding has to last long enough to sweep away the entire crest. If this happens, a breach is almost inevitable and occurs rather quickly. The opening of the breach then releases a large volume of water into the plain.

Chapter 5 will describe all the parameters that can limit erosion (consistent crest profile, cohesion, compaction, regular grass coverage, a gentle landside slope, etc.). However, this does not guarantee complete resistance to overflow erosion.

Internal erosion also leads to equally damaging breaches. Other mechanisms (scouring, sliding) usually do not cause a complete breach, but may initiate a process that will lead to a total breach. The collapse of the slope may lead to internal erosion because of an increase in the hydraulic gradient.

Breaches of the Loire levees in the 19th century, on the Aude and Agly Rivers in November 1999, on the Gard and Vidourle Rivers in September 2002, on the

Rhône River near Arles in late 2003, and the Aude River in Cuxac-d'Aude in November 2005 have shown the limits to the protection levees provide and the potential hazards they can cause. The financial losses are almost always tremendous. The suddenness of these processes may result in the loss of human life. A levee failure in Aramon killed five people in September 2002. The Loire River had three major floods in 1846, 1856, and 1866, causing many breaches and victims. For example, the town of La Chapelle-sur-Loire abutting the levee was destroyed in 1856. As a result of the same flood, the 6-m-high levee in Jargeau completely breached in only 4 hours and released a 2,400 m^3/s flow and 230 million m^3 of water in less than 2 days, creating an 18.7-m-deep erosion pit.

A levee offers protection against medium-sized floods if it is properly built. But it presents a danger during high floods if no specific measures are taken.

Most of the structures in France are earthen levees, including the ones mentioned above. Those made of masonry or concrete resist overflow most effectively when they have a proper foundation. But they can suddenly be broken when sliding occurs or be overturned by water pressure.

Dams almost always have spillways to prevent overflow over their crests, even for exceptionally rare floods. **Can this technique be systematically applied to levees, and more specifically to earthen levees?**

Typology of spillways on levees

Definition of a spillway on a levee

A spillway transfers water from one area to another. If it diverts water from the river to the leveed floodplain, it is a diversion spillway; the flow in the river bed decreases from upstream to downstream of the spillway. If it diverts water from the leveed floodplain into the river, which happens less frequently, it is a return spillway (*reversoir* in French); the flow in the river increases from upstream to downstream of the spillway.

A spillway on a levee is typically a long notch covered with overflow-resistant material such as masonry, concrete, gabions, or concrete riprap (Photos 1.4 to 1.11, Figure 1.9). In some cases, the spillway may be a gravity structure abutted onto earthen levees. Dikes (masonry or concrete levees) can also have spillways. A dike without any notches will not have a spillway, even though its entire surface is resistant to overflow. It functions as its own spillway of sorts, which is of little interest in hydraulic terms. The spillway is connected to both sides of the levee by vertical sidewalls (Photos 1.7, 1.8 and 1.11) or ramps to allow continuous access for maintenance purposes, or even a public road (Photos 1.5 and 1.6).

In France, these kinds of structures are frequently referred to as "overflow levees" or "submersible levees", which implies they resist overflows. To be more specific, we think the correct term for these structures is "overflow-resistant levees".

The term submersible levee implies that other levees are not submersible, which could lead to confusion around the idea that some levees may never be overtopped. Of course, designers who use this wording are implying that levees are "non-submersible up to a certain flood level". But when addressing the complicated topic of flood management, it is best to avoid any confusion. Beyond clarifying the terminology, we have also noticed the term overflow-resistant levee covers two different ideas, depending on the developer:

– For some, this term is simply a synonym of "spillway". They prefer it for psychological reasons. Though this term has its merits, it does not convey the idea of a water intake area.

– For others, spillways and overflow-resistant levees refer to two different and complementary concepts. A spillway is the structure we defined above, with a specific length and two lateral vertical or tilted sidewalls. An overflow-resistant levee is a levee that can withstand overflows without damage thanks to the presence of a water cushion on the landside slope when discharge begins. The spillway creates this water cushion. Spillways therefore make overflow-resistant levees even more resistant to erosion, increasing their efficacy and reducing their cost. Spillways are shorter and built lower than overflow-resistant levees. They can withstand a higher hydraulic load. They are more frequently activated than overflow-resistant levees for three reasons: more frequent overflows, higher hydraulic loads, and the absence of a downstream water cushion when overflow starts.

We will distinguish between these two structures in the rest of the handbook, and will further examine both types in Chapter 3 and Appendix 2 on the Rhône River development.

A diversion spillway is a section of the levee that resists overflow and will funnel the initial discharge. This type of spillway is the main focus of this handbook.

An overflow-resistant levee is a levee section that can withstand overflow without damage. The protection it offers mainly comes from a water cushion provided by a spillway before the overflow starts spilling over the levee.

A return spillway is also a section of the levee that is resistant to flows but discharges water in the opposite direction. It is much less common.

Photo 1.4. The 715-m-long Jargeau Spillway with its 1.5-m-high fuse plug covered with grass. It was built on the left bank of the Loire River two-thirds upstream of the Orléans leveed area from 1878 to 1882 at the site of the 1856 breach. (Photo: DREAL Centre)

Photo 1.5. The Montlivaut spillway on the left bank of the Loire River, which spills floodwater into the upstream part of the Blois leveed area. It was built just like the Jargeau spillway but is covered by a road and located at the site of a breach caused by the 1846 flood. (Source: Gérard Degoutte)

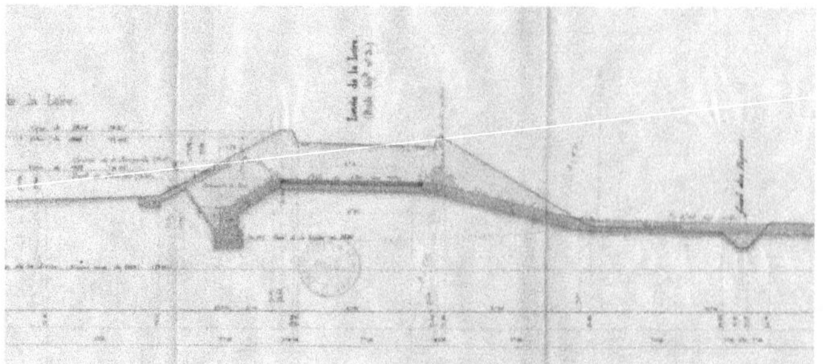

Figure 1.9. Cross-section from December 1886 of the same Montlivaut spillway as in Photo 1.5. The top of the fuse plug is 1.60 m below the 1866 flood level and 1.28 m below the 1856 flood level.

Photo 1.6. The Dampierre Spillway on the left bank of the Loire River, without a fuse plug. The hydraulic function of this structure is not apparent. (Photo: DREAL Centre)

Photo 1.7. One of ten Pitot spillways on the left bank of the Vidourle River (after the flood of 2002 and before the levee was reinforced). The Vidourle is to the left. This spillway was renovated in the 20th century (concrete slabs). (Photo: Gérard Degoutte)

Photo 1.8. Another Pitot spillway on the Vidourle after the levee was reinforced following the 2002 flood. The Vidourle is on the right. We can see the vertical sidewall between the spillway and the regular section of the levee. (Photo: Gérard Degoutte)

Photos 1.9 and **1.10.** Masonry spillway on the right bank of the Giessen River (Bas-Rhin), around 90 m long. (Photos: Bas-Rhin DDT)

Photo 1.11. The so-called diverter spillway in Lattes on the left bank of the Lez River, under construction. This 150-m-long spillway diverts floodwater towards a channel (see Figure 1.11). (Photo: BRL-i)

Typology of diversion spillways according to the hydraulics of the retention area

This section will cover three very different configurations. For now, we will only address the spillway's physical aspects, regardless of whether there are stakes upstream or downstream. The configuration developers adopt is often based on geographical considerations (topography, morphology).

Diversion towards a drainage channel

An interesting situation occurs during flooding when excess flow can be diverted from the leveed riverbed towards another flow channel. The diversion spillway on the levee serves as a flow diverter.

These kinds of configurations already existed on the Loire River in the 16th century and were called *déchargeoirs* (bypass spillways). In these cases, the channel fed back into the same river downstream. However, this type of structure is uncommon; there are just two on the Loire River that we will examine in Chapter 2.

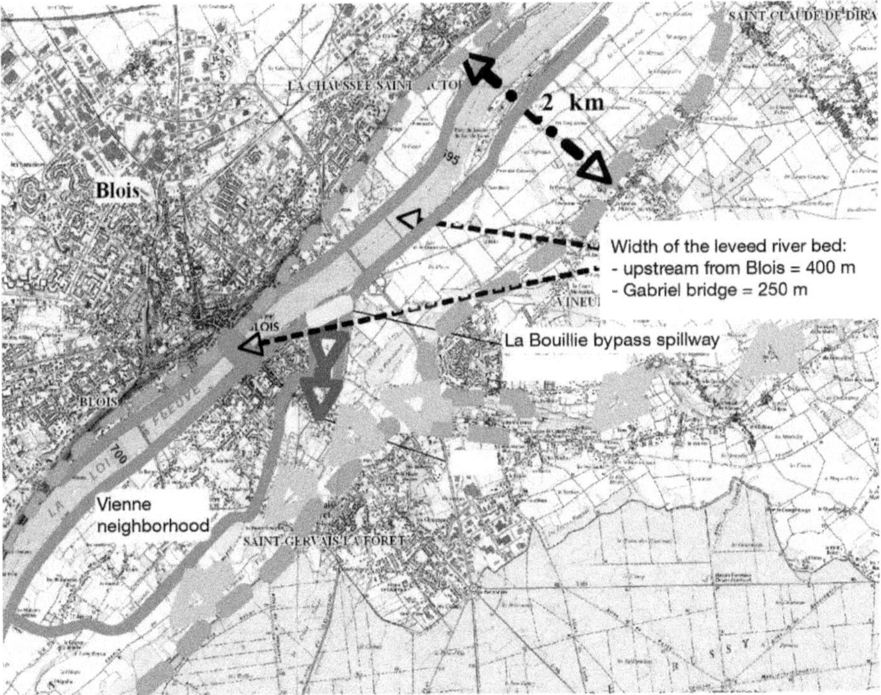

Figure 1.10. The 450-m-long La Bouillie bypass spillway in Blois (yellow line) diverting water into a leveed channel (dark blue arrows) that links up with the Cosson (light-blue arrows), a tributary of the Loire River. The Cosson tributary feeds into the Blois leveed area; for larger than 70-year floods, floodwater from the Loire also spills in via the La Bouillie bypass spillway; for 170-year floods, water from upstream of the Blois leveed area, which is also fed by the Montlivaut spillway, will come in the next day. The Blois leveed area is open downstream. The Vienne neighbourhood is a protected area between the Loire River levee (red) and the southern levee that runs along the La Bouillie channel and then the Cosson tributary and which closes onto the Loire levee. (Source: DREAL Centre)

We have some information about one of the two remaining bypass spillways on the Loire River: the La Bouillie bypass spillway in Blois. It was built in the 16th century, maybe even earlier, to protect a section of the Loire River narrowed by the Gabriel Bridge in Blois. It still serves this protective purpose by supplying floodwater into a channel built in the Middle Ages that connects to the Cosson (a small tributary of the Loire River that drains water out of the Blois leveed area back into the Loire River – Figure 1.10).

We might expect the channel receiving the spillway waters to be uninhabited. Unfortunately, that is no longer the case.

The same configuration was recently adopted on the Lez River in Lattes, downstream from Montpellier. A diverter spillway built in 2008 releases into a flood channel designed to connect with another river, the Lironde. The only difference is that the Lironde River is not a tributary of the Lez River[3] since it flows into the Méjean coastal pond (Figure 1.11 and Photo 1.11).

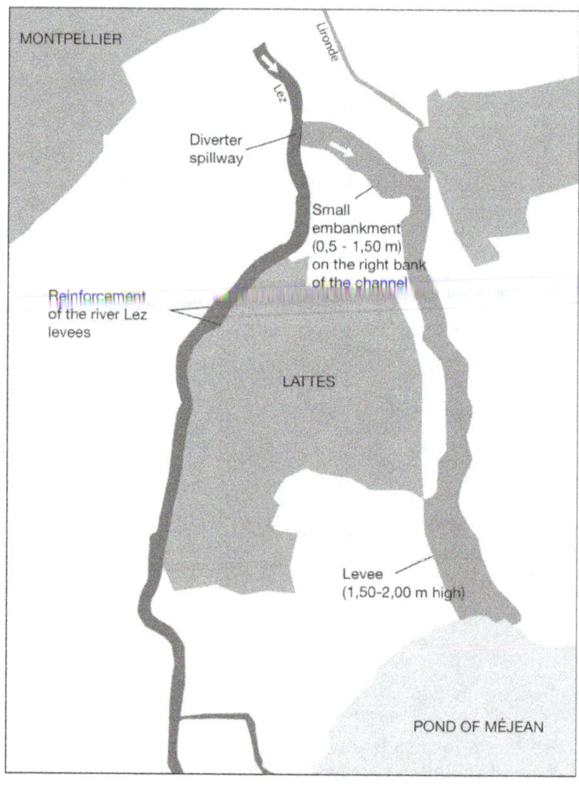

Figure 1.11. The diverter spillway was built on the Lez levee upstream of the town of Lattes (Hérault); it spills into a dug-in channel (light-blue dotted arrow) that connects to the Lironde River. (Source: Montpellier Agglomération, the owner of these structures)

---

3. The Lironde River, sometimes called "ancient Lez", used to be a branch of the Lez delta.

On the Agly River, there were plans in the 2000s to build a large 550 m diverter spillway and a 6.5 km channel that would run along the river until it connected to another watercourse, the Bourdigoul. The project was dropped for financial reasons (Goutx et al., 2004).

The retention channel may end up in a large flood expansion area, as with the Giessen River. A spillway (shown in Photos 1.9 and 1.10) diverts some floodwater to protect the small town of Ebersheim (Figure 1.12). The leveed channel is actually farmland in which two levees connecting to the Giessen River levee were built on both sides of the spillway.

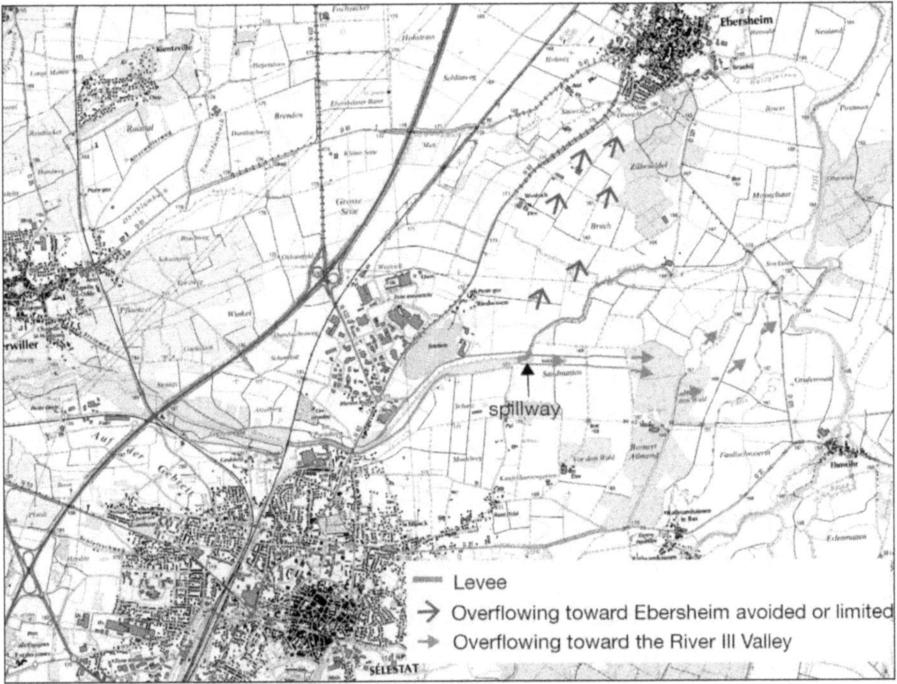

Figure 1.12. The Giessen spillway diverts water into a leveed channel and then into the Ill River alluvial plain. (Source: Bas-Rhin DDT)

To conclude our description of spillways that flow into flood channels, we'd like to present one of the biggest spillways in the world: the Bonnet-Carré spillway on the Mississippi River. It protects New Orleans by diverting part of the Mississippi River's floodwater into a leveed channel, then into Lake Pontchartrain and indirectly into the Gulf of Mexico, bypassing New Orleans (Figure 1.13). It was completed in 1931.

This 2,150-m-long spillway has 350 bays and 7,000 vertical wooden "needles" ranging from 3 m to 3.7 m in height, operated by cranes (Photos 1.12). It has a capacity of 7,100 m^3/s and discharges every ten years on average. It also reduces pressure on the downstream levees.

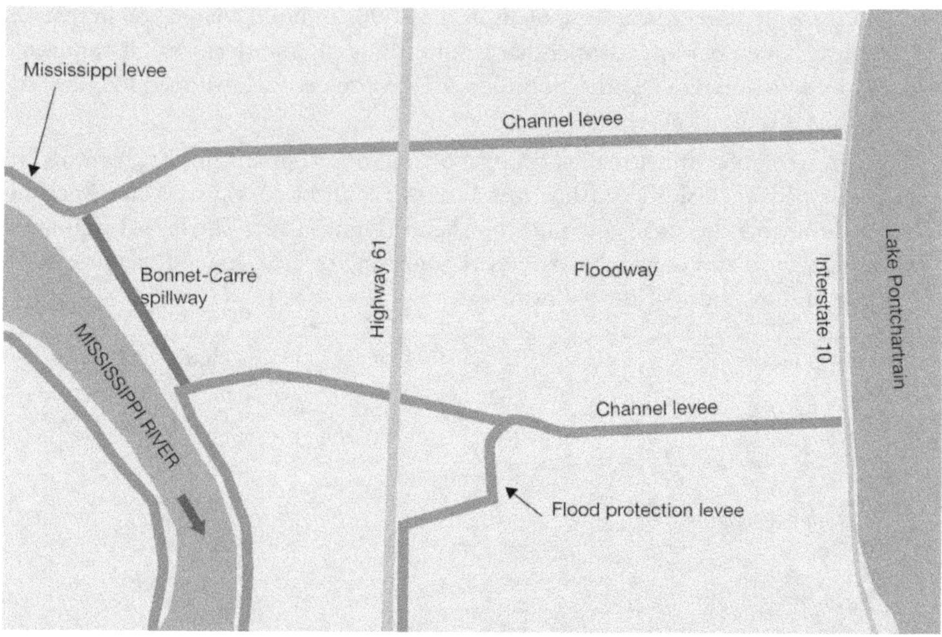

Figure 1.13. Position of the Bonnet-Carré spillway. (Source: US Army Corps of Engineers)

Photo 1.12. Aerial view of the Bonnet-Carré spillway in operation, on the left bank of the Mississippi River. The sticker shows the machine used to move the wooden needles. (Courtesy of US Army Corps of Engineers, public domain)

However, in some or many cases, the area's topography and occupancy may prevent the use of a diversion channel, requiring other configurations.

Diversion towards a storage area

Another configuration involves diverting floodwater into an area where water is stored during flood peak and then returned to the river. Unlike the previous case, water does not flow into the area fed by the spillway. It is just temporary storage. This area is either a controlled flood expansion area or a leveed area.

A typical example is a short retention area with no slope from upstream to downstream. Wherever the spillway is, no water flows downstream, although there may be an initial flow towards the hillside if the river is perched.

Another example is a long retention area fed by a spillway located downstream, as shown in Figure 1.14. The discharge will gradually and slowly head upstream.

In both cases, whether the retention area is short or long, it stores water temporarily without significant downstream flows during flooding. Once the flood is over, the flooded area will drain more or less quickly, if possible due to gravity.

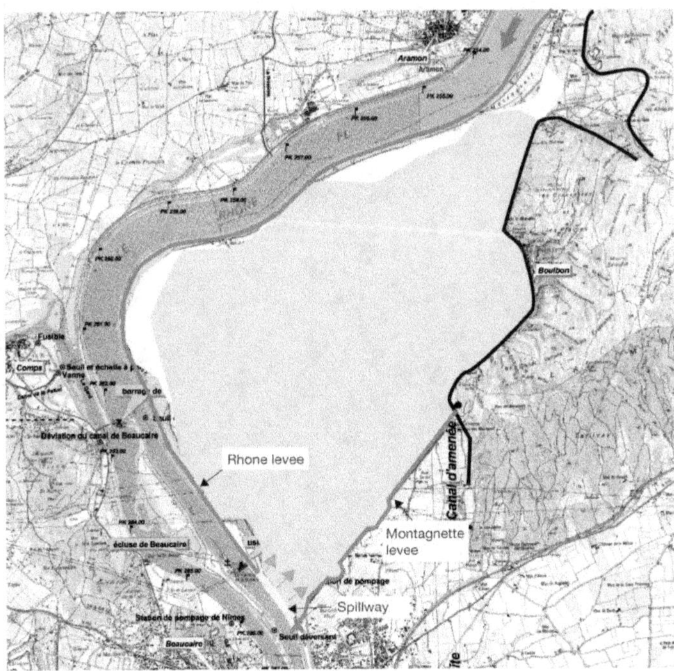

Figure 1.14. The Vallabrègues flood expansion area, on the Rhône River. The left bank, developed in 1969, rests on the south-east side of the ancient La Montagnette levee, which protects the city of Tarascon. The flood expansion area is fed by a 500-m-long spillway with a crest at 10.45 m NGF on the left bank of the power plant's exit channel. Water is discharged at approximately 8,500 m³/s (10 years). At 10,500 m³/s (50 years), the water level reaches 11.7 m everywhere (pink). At 14,160 m³/s (1,000 years), the water level reaches 13.2 m (pink and white). (Source: CNR for SYMADREM on an IGN topographic map)

Diversion towards a flow and storage area

This is an intermediate configuration. The overflow area is initially the source of discharge, and then a downstream obstacle causes the area to be filled from downstream (Figure 1.15).

This configuration is typical of large sections of the Loire leveed areas equipped with Comoy spillways, such as the Ouzouer leveed area where the spillway is completely upstream and the Orléans leveed area where the Jargeau spillway is located two-thirds upstream.

We can see that a long retention area is designed to work as a storage area if the spillway is located downstream, and as a flow and storage area if the spillway is located upstream, as on the Loire River.

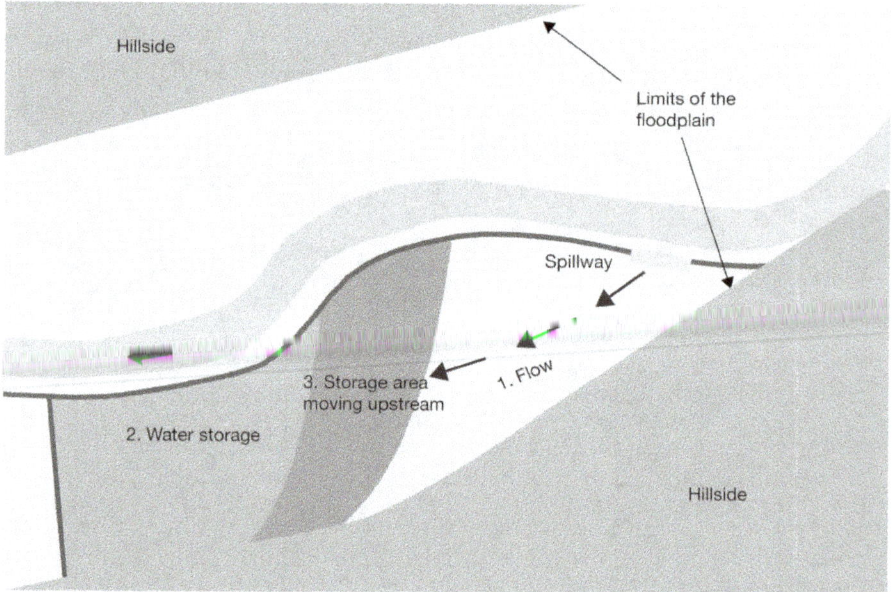

Figure 1.15. Flow and storage area.

The retention area in Figure 1.15 is closed downstream by a transverse levee. However, an area that opens downstream could also function in the same way. Downstream loading is not caused by a levee but by the backwater effect in this non-leveed area. This is typical of the Ouzouer leveed area on the right bank of the Loire River. During strong floods, the following events occur sequentially:
– Water is stored in an expansion area downstream as if there were no levee.
– Water flows from the spillway upstream down the leveed area.
– Storage increases downstream of the leveed area.
– When the floodwater level of the Loire River recedes, the spillway stops discharging.

– Lastly, gravity releases water from storage in the downstream part of the leveed area towards the Loire riverbed.

Lastly, we present the example of the left bank of the Vidourle River, downstream of the Pitot spillways. These spillways discharge into a floodplain common to the Vidourle and Vistre Rivers (Figure 1.16). This floodplain is drained by the Cubelle, a small tributary of the Vistre River. As soon as the flow rate of the Pitot spillways exceeds 10 m^3/s, the Cubelle River begins to overflow and gradually spills into the Vidourle-Vistre floodplain. It cannot drain towards the Vidourle River or the Vistre canal since they are leveed, and the hydraulic load comes from downstream since the canal from the Rhône River to Sète is also leveed. This is a perfect example of the diagram in Figure 1.15, though its flow is more complex because of the many roads or railway embankments and the distance between the spillways and the downstream end of the protected area (around 15 km). Drainage then occurs through various orifices or via pumping and even evaporation.

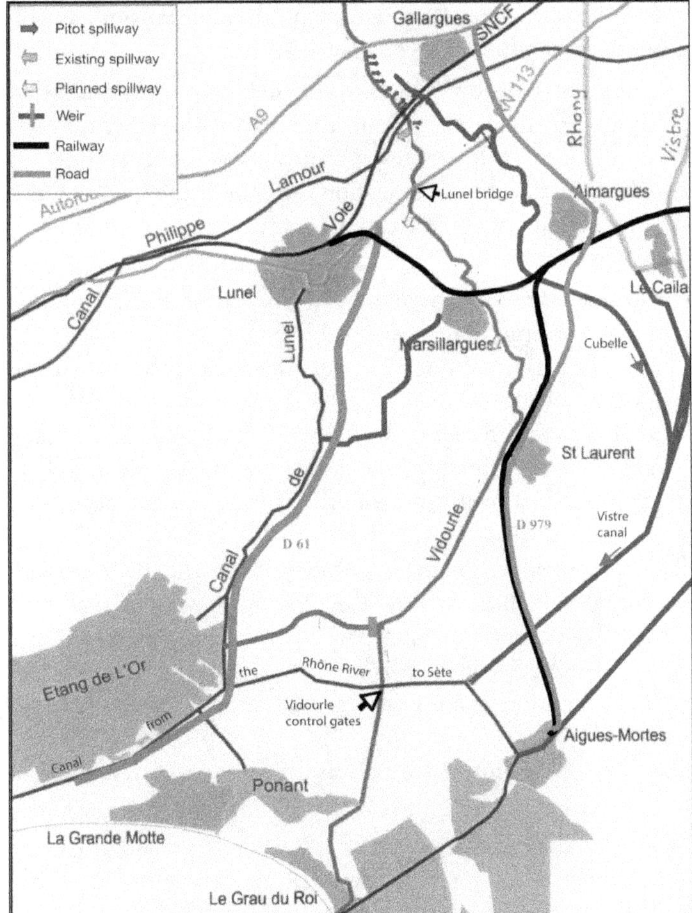

Figure 1.16. Water that passes through the Pitot spillways on the left bank of the Vidourle River flows through the leveed floodplain without being able to return to the Vidourle. It reaches the Vistre Canal with great difficulty since it is leveed and is no longer a coastal river now that it connects to the canal from the Rhône River to Sète. A new spillway was built on the right bank in 2009, and two more are planned.

Typology of spillways according to the purpose of the retention area

Spillways can be found in protected areas and controlled flood expansion areas. In a protected area in the strictest sense, the spillway secures the levee in the area. It is not meant to attenuate the flows. In a leveed flood expansion area, the spillway's purpose is to let through floods that have no impact downstream and reduce the level of stronger floods to benefit the downstream area.

We can see that the role played by the area, whether a protected area strictly speaking, a flood expansion area, or an area that is both protected and serves a flood expansion purpose, will determine the spillway's design.

To distinguish between these two cases, we will use the term **safety spillway** to refer to spillways in protected areas strictly speaking since they are designed to protect the levee. We will use the term **diversion spillways** for spillways in flood expansion areas since they divert a volume of floodwater towards a temporary storage area. Safety spillways in protected areas can serve as diversion spillways, but they are not designed to divert a significant flow of water to attenuate downstream flooding. In areas that offer both protection and flood expansion, a spillway plays both roles: flow diversion and levee protection.

Diversion spillways in flood expansion areas should preferably feed areas that either store water or allow it to flow. Their main role is to reduce the intensity of flooding to protect stakes farther downstream. But this also benefits the levee on the opposite bank, the levees on both banks downstream of the spillway and even levees a certain distance upstream via the backwater effect (if the flow is subcritical).

Safety spillways in protected areas discharge water from strong floods towards protected areas. They have multiple purposes:
– Ensure controlled, slow, and gradual flooding, versus a breach that can occur suddenly and in unexpected locations.
– Limit the volume of floodwater released in comparison to a breach (which typically extends to the levee toe with stronger discharge and longer duration).
– Allow designers to select the first overflow area to minimise impacts according to land use.
– Wherever possible, allow a water cushion to absorb flows and their damaging effects during extreme floods that overflow the crest on the rest of the levee.
– Reduce the frequency of overflows on other parts of the levee (upstream and downstream) and the opposite levee, if there is one.
– Serve as a warning to facilitate decisions to evacuate an area or recommend safety measures.
– Maintain awareness of risks by eliminating a false sense of security that can be dangerous.

The main difference between these two types of spillways is the developer or designer's goal: to secure the levee (taking the stakes into account) or to attenuate floods. It is harder to differentiate between their physical aspects:

– A safety spillway also diverts flow (towards the protected area), but we will see that this is usually minimal in proportion to the river's flow.
– A diversion spillway that diverts water towards a flood expansion area will lower the waterline downstream and even a certain distance upstream in the case of subcritical flow. Often, the river is also leveed on the opposite bank and both banks downstream. This means that lowering the waterline bolsters the safety of these other levees but also of the flood expansion area itself.

Lastly, rather than feed into an open or closed area where the flow will spread out, the diversion spillway can feed into a natural or built channel that concentrates the flow. This channel can connect to the same watercourse farther downstream, a tributary, another watercourse, or the sea.

Photos 1.4 to 1.6 show a safety spillway designed to protect a leveed area on the Loire River. Photos 1.7 and 1.8 show diversion spillways diverting water towards a floodplain. Photos 1.9 and 1.10 show a diversion spillway diverting water towards a flood expansion area. Photos 1.11 and 1.12 show a spillway diverting water towards a flood channel. But this difference is not visible in the photos.

In summary

- A diversion spillway can release floodwater into a flood expansion area that stores the water until the flood is over, then returns it to the river.
- It may feed a flood channel that transfers water much farther downstream into the same watercourse, another watercourse, a large lake, or the sea. There is no protected area or flood expansion area.
- A safety spillway protects the levee in a protected area.

Typology of return spillways (*reversoirs*)

There is no need for a return spillway in a short flood expansion area, short protected area, or in the case of spillways built downstream of these areas. It is only when the spillway is built upstream of a long area that a return spillway is useful. It must have the same discharge capacity as the spillway upstream, or less if the volume of water stored in the area before discharge is significant compared to the volume of floodwater diverted into the area. This is typically the case for flood expansion areas.

Return from a drainage channel

In this configuration, a flood channel from upstream flows back into the river, crossing the levee through a structure with free overflow. This structure must return flows diverted upstream and prevent backward erosion into the channel. It can also prevent the channel from being filled with water from downstream before it is filled from upstream.

Return from an expansion area or a protected area

In this configuration, the floodwater retention area (flood expansion area or protected area) has a slope and there is a risk of overflow over the downstream part of the levee towards the river. A return spillway can prevent or limit these overflows (Figure 1.17).

This type of structure can also work in both directions (as a diversion spillway or return spillway) according to the type of flood. It can work as a diversion spillway for very sharp floods which have not yet filled the leveed area with much water. However, for longer floods, the retention area may be full and "return" towards the river even if the river level is not or no longer above the spillway.

It may have a fuse plug, as in Vouvray on the Loire River near the junction with the Cisse River.

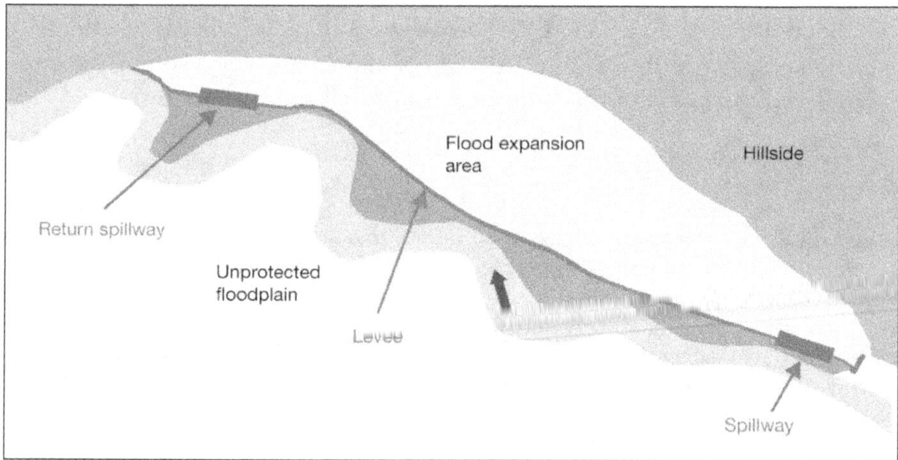

Figure 1.17. Flood expansion area with a spillway and return spillway.

Review of overflow areas

In physical terms, a levee spillway can supply water into three types of areas:
- A flow area (or channel)
- A storage area
- A flow area (upstream) then a storage area (downstream)

In terms of the decision maker's objective, these areas may be water retention areas (flood expansion areas or channels) or on the contrary have important stakes that requiring keeping out as much water as possible (protected area). This leads to the six combinations below.

Table 1.1. Overflow areas

	Water retention area	Area with stakes
Flow	Potentially leveed flood channel that may connect to the same watercourse farther downstream, another watercourse, a pond, or the sea	Floodplain in a perched river with some stakes (farmland, scattered constructions, roads)
	Examples: the Lez River in Lattes, the Mississippi River upstream of New Orleans, the Loire River at La Bouillie (in the Middle Ages)	Examples: the Pitot spillways on the Vidourle River, spillways on the Aude River between Cuxac and Coursan, the Loire River at La Bouillie (now)
Storage	Flood expansion area with downstream diversion spillway	Protected area with downstream safety spillway
	Example: Vallabrègues flood expansion area (Rhône River)	Example: Aramon protected area
Flow and storage	Flood expansion area with an upstream diversion spillway	Protected area with an upstream safety spillway
	Example: the Printegarde flood expansion area (Rhône River)	Example: Ouzouer leveed area (Loire River)

This typology is simplified and the wording "water retention area" is a bit theoretical. For example, in the Vallabrègues flood expansion area, there are many inhabited areas on high ground. And in some areas considered to have stakes, they can be very scattered and mostly located on high grounds (see Vidourle).

Chapter 3 will cover spillway design, which varies greatly depending on whether the configuration is a diversion channel, a flood expansion area with limited stakes, or a protected area with important stakes.

2

The historical background of spillways in France

Gérard Degoutte
Jean Maurin

The construction of spillways on river levees is not a new concept. This chapter will provide a non-exhaustive historical background for France.

1584 to 1891: The first spillway constructions on the Loire and Vidourle Rivers

1584-1867

The oldest known spillway in France seems to be the La Bouillie spillway in Blois. It is on the left bank of the Loire River, upstream of the town centre. In the first available records dating from 1584, this spillway was called a *déchargeoir* (bypass spillway). It was initially designed as a simple gap in the levee that was extended by the Blois Canal and used to divert floodwaters towards the Cosson River, a tributary of the Loire River (Figure 1.10). Spillways as specific structures were only built after a series of disastrous floods, including on the Pentecost in 1733. In 1740, engineer Louis de Règemorte oversaw the construction of a paved path that connected to the levees via ramps. After several flood events, people living in the leveed area managed to have it removed in 1785 by having a levee built in front of the bypass spillway. In January 1789, an ice breakup swept away a major part of this levee. However, it may also have been opened during the night by residents of the opposite bank in Blois.

In 1791, the bypass spillway was rebuilt and heightened by 1.4 m at the suggestion of the Ponts et Chaussées (Civil Engineering Authority), despite opposition from people living in the slums on the right bank of the river in Blois. Then came new floods, new damage, and new constructions. In 1867, following the exceptionally heavy flood of 1866, the levee was reinforced and extended. Small earthen ridges were later created on both sides of the road. The spillway performed properly in 1907, for the last time. All this information comes from a historical overview commissioned by the DREAL Centre-Val de Loire (Regional Directorate for Environment, Development, and Housing in the Centre-Val de Loire Region) and the DDT in Loir-et-Cher (Departmental Territory Directorate in Loir-et-Cher) (De Person, 2001).

Other bypass spillways were built. Following the floods of 1707, 1709, 1710, and 1711, Louis XIV's government drafted a rehabilitation plan in 1711. This plan involved filling in breaches that had been opened by previous floods, raising the levees, and building bypass spillways where none existed. A construction programme between Gien and Tours led to the installation of many bypass spillways in the narrowest sections of the leveed riverbed. There are records of six other similar programmes. However, the flood on the Pentecost of 1733 opened many breaches and destroyed the bypass spillways. In response to pressure from local communities and engineers' doubts about how to maintain these structures, the royal government had all the bypass spillways removed starting in 1733. Only those built before 1711 were spared: La Bouillie upstream of the town of Blois and Saint-Martin-sur-Ocre upstream of the town of Giens. Both these bypass spillways are still in place.

The Pitot spillways

Apart from the bypass spillways on the Loire River, the spillways on the Vidourle River seem to be the oldest. After the particularly disastrous flooding of the Vidourle River in November 1754, the États du Languedoc (States of Languedoc) asked renowned engineer Henry Pitot to find a solution. Starting in 1764, Pitot oversaw the construction of ten masonry spillways on the left bank levee of the Vidourle River at Gallargues and in front of the Ambrussum oppidum, downstream of what is now the A9 motorway (Photos 1.7 and 1.8). They were completed in 1773. The challenge was to leave the 1754 flood breaches in place while preventing them from extending or broadening (Cœur, 2008). The story goes that in 1776, the town of Gallargues requested that the entire levee be lowered to the level of the bypass spillways, only to call for the spillways to be closed in 1836.[4]

These spillways are diversion spillways. They are installed about one metre below the levee crest and run 20 m long. Despite their limited length, these spillways play an essential role in the entire Vidourle River plain downstream of the A9 motorway. The first spillway goes into effect at a flow of 710 m^3/s at the A9, and the last one at 970 m^3/s. However, all of the spillways will be heavily activated at 1,000 m^3/s, that is peak flow for a 10-year return period. Above that level, the Vidourle River's flow will keep increasing at a slower pace. For example, in September 2002, for a 400-year return period flood, the incoming flow was 2,330 m^3/s, which the spillways lowered to 1,665 m^3/s.

---

4. Service hydraulique du Gard (Gard Hydraulics Service), Agriculture Department, 1931, town of Gallargues, levee maintenance, Chief Engineer's report requesting a 12,000-franc subsidy.

1867-1891

Back to the Loire River, one hundred years after the Pitot spillways. The Ponts et Chaussées engineer Guillaume Comoy demonstrated the harmful effect of repeatedly raising levee crest levels under the illusory hope of rendering them "non-submersible". This raising only increased the maximum flow each time, and Comoy stated "the scourge keeps worsening". In 1867, he developed a plan to build 20 spillways on 18 out of 33 leveed areas (see Appendix 2). These spillways were designed to let water flow into the leveed areas during major flood events to limit the pressure on the downstream part of the leveed area. Due to economic difficulties during the 1870 post-war period and opposition from local communities, only eight spillways were built between 1867 and 1891. The typical configuration was an upstream spillway that sent water from strong floods into leveed areas that were open downstream. They were very long masonry spillways (Photos 1.4 to 1.6, Figure 1.9). The weir level was sometimes raised by a 1.35- to 1.75-m-high embankment meant to serve as a fuse plug. The Guétin spillway, in the Bec d'Allier Valley, is the only spillway to have functioned, just once in 1907.

20th century: Several scattered constructions and major development of the Rhône River

1917. Still, on the Loire River, two new spillways that were not part of the initial plan in 1867 were built farther downstream in the Divatte leveed area (Loire-Atlantique department[5]).

1952. On the Aude River, the 1952 flood caused many levee breaches, leading to the construction of concrete spillways at the breach locations. Three concrete spillways were installed on the left bank of the Aude River between Cuxac (upstream) and Coursan (downstream): Horo de Blazy (186 m) 1952[6] (308 m), and Prat del Raïs (350 m). These spillways were effective during the 1996 and 1999 floods. In 1996, the incoming flow rate in Cuxac was 1,230 m^3/s, which decreased to 830 m^3/s after the first spillway, then 720 m^3/s after the second, and 600 m^3/s after the third. In 1999, though the flood was much stronger than in 1996, the flow that released into the Aude River downstream of the last spillway was barely higher: 1,970 m^3/s in Cuxac, 1,030 m^3/s after the first spillway, 810 m^3/s after the second, and 670 m^3/s after the third.[7]

1962. During construction of an 8 km levee on the Reyran River in Fréjus that extended to the sea, a 210 m spillway was built on the right bank, one kilometre

---

5. France is divided into eighteen administrative regions. Regions are further subdivided into two to thirteen administrative departments

6. This is the actual name of the spillway, which serves as a historical reference.

7. Source: BRL-i, October 2001, Development on the low-lying floodplains of the Aude River - Stage 1C, hydrological and hydraulic study.

from the upstream end. The levee had an earthen fuse plug and concrete slabs on the upstream face of the fuse plug that extended from the concrete slabs on the entire levee (Photo 5.16). The fuse plug was designed to erode and overturn the concrete slabs in the event of overflow. The fuse plug crest was 50 cm below the levee crest. The spillway has yet to be used, and a 2011 study suggests that it would only serve in the event of a 500-year flood.[8]

On the Rhône River

As part of the development of the Rhône River, many levees were installed along its entire length. Flooding in certain areas was preserved, or even optimised by several spillways and structures with gates, siphons or pumping stations. In all the structures on the Rhône River operated by the Compagnie Nationale du Rhône (CNV), there are 18 spillways and long sections of overflow-resistant levees. These structures feature a variety of designs based on the geographical and hydraulic features of each site. Most of them are levees whose crests and slopes are protected both upstream and downstream by riprap, bituminous coatings, or concrete slabs. The overflow flood return periods range from 2 to 20 years. Flooding often comes from downstream through a side canal without any specific overflow structures. Drainage consistently occurs when the floodwater subsides, either by gravity or pumping. A detailed description of these structures can be found in Appendix 3.

21st century: A relative acceleration in construction

2003. The levee Pitot built in 1750 to protect the small town of Aramon (at the junction of the Gardon and Rhône Rivers) gave way during the flood of the Gardon River in September 2002. It was rebuilt in 2003 with a 900 m spillway at the same level as the former levee. In the overflowing area, the levee was raised by 1.10 m. The spillway is made of 200–400 mm rockfill concrete and covered in topsoil. This quick intervention paid off. As the structure was nearing completion, it was tested by a flood of the Rhône and Gardon Rivers in December 2003, with a moderately high (about 17 cm) overflowing nappe (Mallet et al., 2004). The spillway worked properly (Photo 2.1).

2006. On the Durance River in Les Mées (Alpes de Haute-Provence department), the Annonciade levee was repaired by the Syndicat Mixte d'aménagement de la vallée de la Durance (SMAVD – Joint Union for Joint Union for Development in the Durance River Valley) and a 120 m spillway was added.

2006. The Comps spillway was rebuilt on the right bank of the Gardon River at the confluence of the Rhône River. It replaced the former spillway built around 1970 during the general development of the Rhône River, which had a fuse plug section in the shape of two embankments. During the flood of the Gardon

---

8. Source: Hydratec-Terrasol study for the town of Fréjus.

River in September 2002, the fuse plug was overtopped but did not erode as planned. The new spillway, made of rockfill concrete, no longer has a fuse plug (see Chapter 5).

Photo 2.1. The Aramon spillway the day after the December 2003 flood (the riverside is on the left). (Source: DDT 30)

2008. On the Lez River, a so-called diverter spillway was built in 2008 upstream of the city of Lattes to partially divert floodwater into another watercourse, the Lironde River (Photo 1.11, Figure 1.11). The spillway is 150 m long, with a 1.8 m height difference between its crest and the levee crest. It is activated for 30-year floods (400 m^3/s) and reduces the flow rate of the Lez River's 100-year flood from 755 m^3/s to 570 m^3/s.

2009. On the Vidourle River, the Syndicat Mixte Interdépartemental d'Aménagement et de mise en valeur du Vidourle (SIAV – Joint Interdepartmental Development Board for the Vidourle River) built the Lunel spillway in 2009 on the right bank of the Vidourle River, slightly downstream of the Pitot spillways (which are all on the opposite bank). This spillway replaced one that was built a bit farther downstream in 2001, which had broken twice previously. The new spillway, designed by ISL, is 500 m long and made of gabions and Reno mattresses, with a concrete beam forming the weir (Figure 5.8).

2009. The levee on the left bank of the Meuse River in Givet gave way during the January 1994 flood, inundating the harbour and parts of the town. In 2009, the town of Givet reinforced and raised the levee and added a 300-m-long spillway with a 1 m difference in the level of its crest and the levee crest. The spillway was installed at a 100-year flood level.

2011. The SMAVD built a 180-m-long spillway in the middle of the Durance River in Rochebrune, downstream of the Serre-Ponçon dam.

2012. The Syndicat Mixte d'aménagement, de gestion et d'entretien des berges de la Seine et de l'Oise (SMSO — Joint Planning, Management, and Maintenance Board for the Seine and Oise Riverbanks) installed five spillways on the left bank of the Seine River in Sartrouville. The spillways were designed by Egis Eau with a cumulative length of 136 m for a 2.5-km-long levee. Because this area is highly urbanised, the spillways were built just 20 cm under the levee crest to fill the protected area so water does not fall more than 50 cm when the level of the Seine approaches the levee crest. This creates a water cushion of around 70 cm.

Other projects have been built recently or are well underway, such as on the right bank of the Vidourle River downstream of Lunel and Marsillargues, on the Isère River upstream of Grenoble, on both banks of the Rhône River between Tarascon and Arles and between Beaucaire and Fourques, and on both banks of the Petit-Rhône River.

In summary

Spillways have been used for centuries on rivers for both slow floods and supercritical flows. Most of these spillways are very long. They were often built to mend breaches soon after they occurred. At times the levees remained unchanged at the breach. Other times the levees were raised and the spillway crest was built at the level of the former levee crest. Or the spillway crest was placed at the same level as the historic flood.

However, many levee systems on rivers do not have any spillways.

In the future, we hope it will become the norm to preventively install spillways rather than wait until catastrophic flooding occurs. The increasing number of decision-makers and designers who understand their importance is a step in the right direction. The French regulation stemming from the 11 December 2007 decree, particularly on risk assessment, should also encourage this preventive action

3

The hydraulic design of spillways

Gérard Degoutte
With contributions from David Goutx and Francis Fruchart

The spillway's impact on the flooded river's waterline

The spillway's impact on waterlines is shown in Figures 3.1 to 3.3 for all potential flow rates, from subcritical to supercritical.

Downstream, regardless of the flow regime, the river's flow will be reduced by the diverted flow and the waterline is lowered over a rather large distance, theoretically all the way to the sea or wherever the diverted flow is restored to the watercourse.

Upstream, if the flow is subcritical, the waterline will be lowered via the backwater effect. This effect will gradually decrease upstream as head losses along the watercourse "dilute" the benefit of the lowered downstream level (Figure 3.1). When the diverted flow is significant and the flow is subcritical but fast, the flow may become supercritical at the level of the weir and form a hydraulic jump (Figure 3.2). This unstable situation should be avoided. It may occur when the weir is long and has a high relative drawdown (low p height relatively to H). With a supercritical flow (Figure 3.3), the waterline will not be lowered upstream as the flow is driven from upstream. In addition, the diverted flow is limited in this case (see Section 3.2). However, this rarely occurs on river spillways.

Of course, where this lowering occurs, the critical moment of levee overflow will be delayed, even for the levee on the opposite side without a spillway.

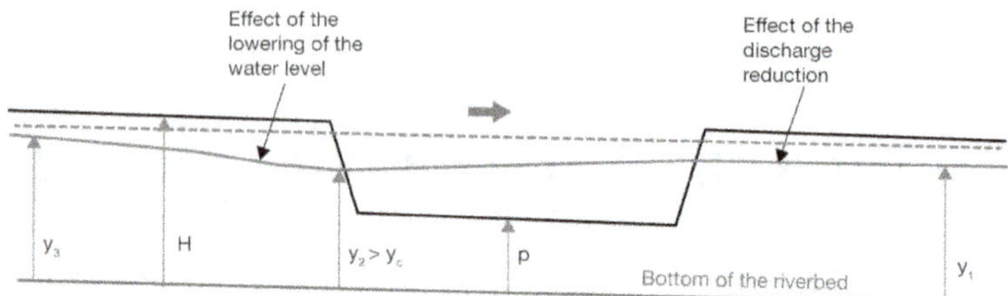

Figure 3.1. Hydraulic effect of a spillway in operation for a flow that starts and remains subcritical. The dotted line is the waterline without a spillway. The solid line is the waterline with a spillway. The water levels y3 and y1 correspond to uniform upstream and downstream flow regimes. yc refers to the critical flow depth. Notes: the variation in level from upstream to downstream is exaggerated in the drawing; the waterline upstream of the spillway is lowered much more gradually.

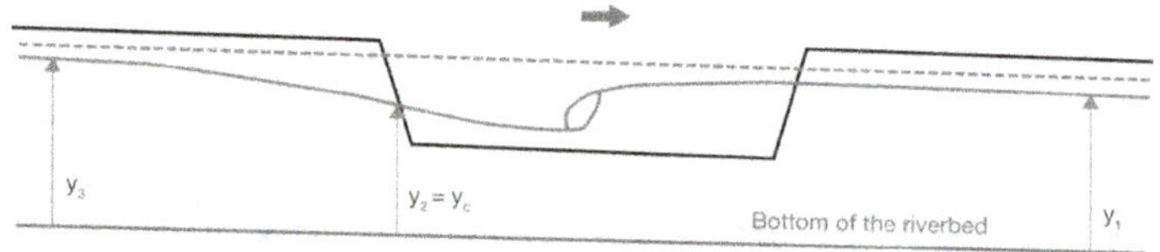

Figure 3.2. Hydraulic effect of a spillway in operation for a subcritical flow that becomes supercritical because the downstream level is lowered. The dotted line is the waterline without a spillway. The solid line is the waterline with a spillway.

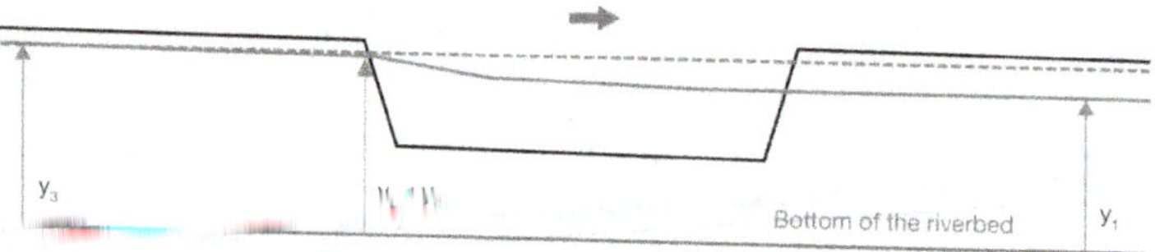

Figure 3.3. Hydraulic effect of a spillway in operation with a consistent supercritical flow. The dotted line is the waterline without a spillway. The solid line is the waterline with a spillway.

Warning: these diagrams imply that the bottom of the riverbed does not change, but it could be increasing or decreasing. Furthermore, the levee itself may cause this change. And even when a leveed river has reached an equilibrium, installing a spillway might alter this balance over time. We will cover this topic in Chapter 4.

Impact of the spillway on the waterline in the protected area

The spillway conveys water into the protected area. If it is a closed area, a water cushion will form downstream as soon as overflow begins and will progress upstream as the flow regime increases and time passes.

On the upper part of Figure 3.4, the levee crest runs parallel to the waterlines during floods. Flood ③ will provoke a generalised overflow that will encounter a water cushion only in the downstream part of the area. The levee is therefore at risk upstream.

On the lower part of Figure 3.4, the levee crest slope is steeper than the waterlines during flooding. Flood ③ will provoke overflow only in the downstream section, where a water cushion has already formed. With a stronger flood, water will overflow the entire levee crest, but the water cushion will have developed upstream. This means the levee toe is protected at all times thanks to the water coming in through the spillway.

The most important cause of erosion at the levee toe will then be contained. Other potential sources of erosion on the face will be limited if the levee is made of well-compacted cohesive soil with a good grass cover (see Chapter 5). In this case, erosion is less likely and will occur later, with slower kinetics than in the first situation. Ideally, the water level in the plain will reach the level of the levee crest as water gradually flows over the crest upstream (unlike in Figure 3.4). This would require a small protected area and a large spillway, ideally a controlled movable device. The manager would need extensive technical skill to control such a device and may encounter opposition in the protected area (much less so in a flood expansion area). Therefore, this kind of device is not appropriate in all circumstances. In addition, encouraging the formation of a large water cushion will flood more stakes in the area. Each situation needs to be studied individually, considering both hydraulic features (such as slope) and the stakes to be protected.

Let us now examine an open area. In this case, the flow conveyed by the spillway will form a thinner water cushion. The worst possible scenario is that of perched river beds, as water introduced behind the levee will not be concentrated at the

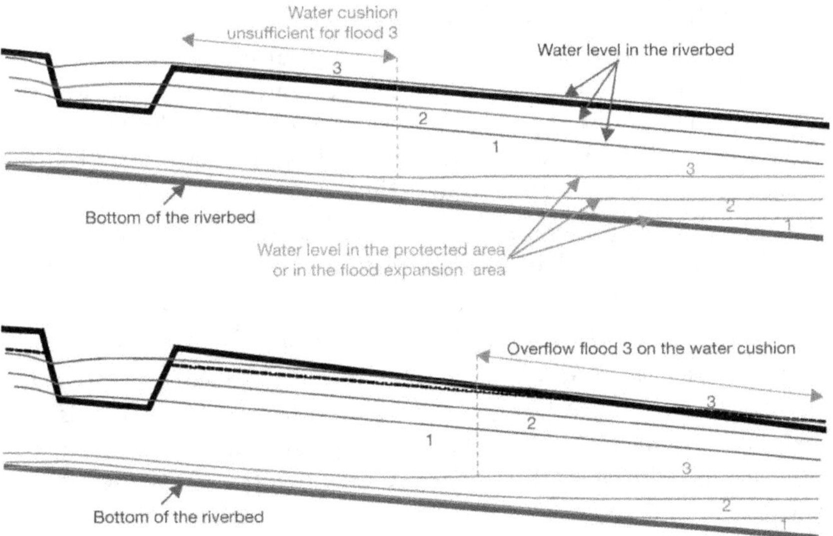

Figure 3.4. Top: longitudinal levee parallel to the waterlines during flooding. For increasing flood levels ① to ③, all higher than the protection level, the closed protected area or flood expansion area will gradually fill in with water from upstream. Flood ③ will overflow consistently over the entire levee up to the spillway, but the water cushion will only form downstream because of the floodplain's slope. Bottom: The levee profile is altered, with a slope higher than that of the waterlines. The same flood ③ will only overflow downstream, where a water cushion has already formed. With an even stronger flood that overflows the entire levee, a bigger water cushion will form everywhere.

levee toe, but at the base of the hillside. In theory, a water cushion cannot be formed unless adaptations are made, such as adding a spur parallel to the levee. It should not be viewed as a secondary mini-levee, but rather as the bank of a channel to guide overflow water. But this is not always possible due to the location. In other words, the spillway may not play its cushioning role in every configuration.

> ## Conclusion on the spillway's stilling role:
>
> It is highly recommended, but not always possible to form a water cushion along the bottom of the levee in the protected area. If possible, a new levee's profile should facilitate progressive overflows from downstream to upstream. This is not a new idea, as our ancestors designed these systems centuries ago, such as on the Saône River. For existing levees that are not designed to create this water cushion, it may be hard to change the profile for financial or regulatory reasons. Even if the formation of a water cushion is preferable, there is still a risk of erosion and breaches. Therefore, we will continue to consider that a danger flood level is one that starts overflowing any part of the levee. This level would be a bit lower than flood ③ in Figure 3.4.
>
> The hydraulic assessment should also consider the importance of the protected stakes and their location in the affected area, both from a plan and an elevation view.

The spillway's flow law

Front weir with a rectangular section

The spillways we are interested in have a lateral flow intake area. But we first want to go over the flow law of front weirs, which are simpler cases since flow runs perpendicular to the structure (Figure 3.5). As a result, the water level and head will remain consistent along the entire length of the sill. We will then show how to modify this law for a lateral spillway.

In the case of a broad-crested weir, the streamlines will run parallel and the flow depth above the sill will be the same as the critical level. The flow law for a weir of length L is expressed by:

$$Q = \mu \cdot L \cdot \sqrt{2g} \cdot H^{3/2} \text{ with } \mu = 2/3\sqrt{3} = 0.385,$$

as long as the weir flow is free, meaning it is not affected by downstream conditions. We can say that the weir controls the flow. $H = h + V^2 / 2g$ is the specific head and may be mistaken for height h in gently sloped rivers, as long as we are far enough upstream of the weir, so we are not in the area where the streamlines curve.

With weirs of a different shape, the same formula is used but the flow coefficient is calculated using tests that are based primarily on the shape of the weir, and to some extent on the head.

The weir flow is submerged when H' is higher than $2H/3$ and the law becomes:

$$Q = \mu' \cdot L \cdot H' \sqrt{2g \, (H - H')} \text{ with } \mu' = 3\sqrt{3}\mu / 2. \quad H = h + V^2 / 2g$$

is always positive if there is discharge and H′ negative or positive refers to specific heads relative to the weir crest (Figure 3.5) (Degoutte, 2006).

It is important to check that the weir is not submerged since, when it is submerged, it no longer serves as a control section and its conveyance will gradually diminish as the tailwater level increases. This will also occur with lateral spillways, the ones that interest us.

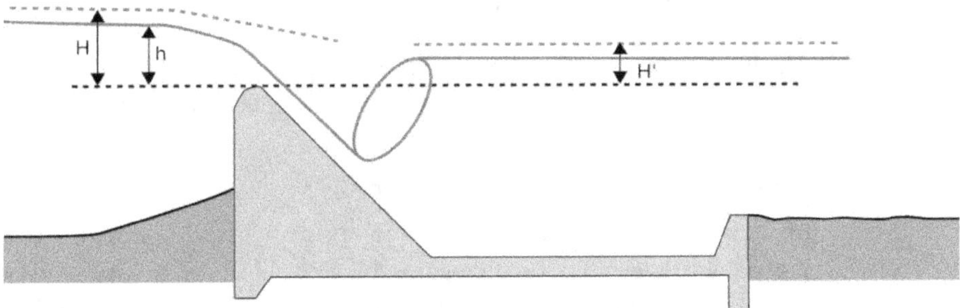

Figure 3.5. Flow over a front weir.

μ is the weir flow coefficient, which ranges from 0.33 to 0.50 depending on the weir's head and whether it has an adequate or inadequate profile. Below is the flow coefficient in order of magnitude for weirs without lateral contraction for low heads (around 50 cm):
– 0.42 for thin weirs
– 0.33 to 0.40 for thick weirs with a sharp-edged rectangular section (0.33 heads less than 40% of the weir's crest width; 0.40 for heads 1.6 times this width)
– 0.36 to 0.45 for thick weirs with a rounded rectangular section (0.36 for heads less than 40% of the weir's crest width; 0.45 for heads 1.6 times this width)
– 0.37 to 0.48 for 20-cm-wide trapezoid weirs with heads of around 20–50 cm
– 0.42 to 0.48 for triangular weirs with a downstream slope below 1:3
– 0.36 to 0.39 for triangular weirs with a downstream slope between 1:5 and 1:10
– about 0.5 for shaped weirs (Creager-type)

(Sources: Lencastre, 1996; Laborie et al., 2005; Degoutte, 2006)

Front weir with a trapezoidal section

Spillways built on levees often have a trapezoidal section to allow monitoring and maintenance vehicles through during non-flood periods. Vertical sidewalls are replaced by two gently sloped ramps (typically 1:3 to 1:5). To our knowledge, there is no formula in the literature that applies to thick trapezoidal weirs. However, we can adapt the rectangular weir formula by changing the proof.

For a front rectangular weir of length L in free flow, we already noted that:

$$Q = S\sqrt{2g(H-h)} = \mu \cdot L\sqrt{2g} \cdot H^{3/2}$$

with S = L·h, the (critical) flow section and h = 2H / 3, where μ = 0.385.

For a trapezoidal weir with a base length L and an average batter m, we use the same calculation and introduce a correcting factor of $\lambda_{tr}q$ based on the head. That is:

$$Q = \lambda_{tr} \cdot \mu \cdot L \cdot \sqrt{2g}H^{3/2} \text{ with:}$$

$$\lambda_{tr} = \left(3 - \frac{4}{1 + \sqrt{1 + m'}}\right)^{3/2} \text{ considering that } m' = m\frac{H}{L}$$

This m' coefficient is a corrected batter due to the slightly denser nature of the spillway. The formula, which is not applicable to L = 0, will produce a result with a less than 1% error margin if m'< 0.28.

For a rectangular section, m = 0, m' = 0 and the result is indeed λ_{tr} = 1.

Below are some correcting factor values for ramps with a 1:5 slope. The effect of the trapezoidal shape may be disregarded for very long structures, which are common. But this is impossible for short structures. Please note that L is the base length regardless of the footprint of both ramps.

m	5	5	5
H	0.5	0.5	0.5
L	10	50	500
m'	0.25	0.050	0.005
λ_{tr}	1.172	1.037	1.004

Lateral spillways

Spillways on levees are lateral. First, unlike front spillways, the overflowing water layer in lateral spillways varies between the upstream and downstream sections of the spillway. The spillway law will therefore result in a flow Q' per metre of overflow on abscissa *x*, that is:

$$Q' = \left|\frac{dQ}{dx}\right|.$$

Of course, if the abscissa is considered positively moving downstream, this value will be negative as the watercourse flow decreases as it moves downstream. Thereafter we will leave out the "minus" sign to express the absolute value of this linear flow.

Furthermore, flow over a lateral spillway will change the angle for streamlines even if the receiving structure is perpendicular to the spillway (Figure 3.5). Compared to front flow, the flow regime will decrease. And for the same water level upstream of the weir, head losses will be higher on a lateral spillway than on a structure with front flow. Consequently, the speed and volume of the flow will be lower on a lateral spillway (irrespective of the effect induced by the change in direction).

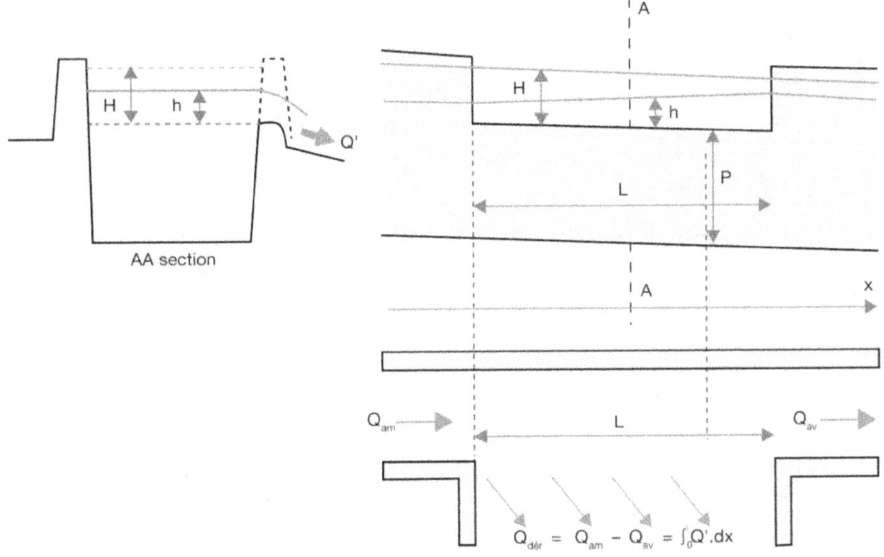

Figure 3.6. River section and profile with defined parameters. Plan view showing the direction of flow.

Work conducted by Hager (1987) allows us to take these effects into account. Hager resolved the question using the theory of gradually changing flows, taking a unidirectional approach based on a uniform distribution of velocity and hydrostatic distribution of pressure. He examined the common occurrence of gentle slopes (<1–2%), which allows us to consider that the head H will remain constant. This leads to a particularly simple situation that is similar to the case of a front spillway:

$$Q' = \left| \frac{dQ}{dx} \right| = \mu \cdot c \sqrt{2g} H^{3/2} \quad ,$$

where correcting factor c is conveyed by:

$$c = \frac{h}{H} \sqrt{\frac{h}{3H - 2h}}$$

The flow head of the watercourse is $H = h + V^2 / 2g$, where $V = Q / S$ and S is the wetted cross-section in the river (stopped crossing over the spillway). All these variables are functions of abscissa x.

The total diverted flow may be calculated by integrating Q' along the spillway (Figure 3.6).

These results have been confirmed by experiments conducted on rectangular canals in the Zürich laboratory, with a maximum predictive error of 5% for the total diverted flow.

We have also assumed that the river takes a prismatic shape along the spillway (neither convergent nor divergent), and we deleted another coefficient that takes this effect into account.

We can see that the corrective term is below 1 and tends to 1 for a velocity that tends to 0. It will be far below 1 for high velocities. Therefore, it would generally be a mistake to ignore the lateral effect.

For trapezoidal flow sections often found on levee spillways, in the absence of references we suggest using the same correcting factor as in the front weir formula (see above).

Once the initial design is complete, verifications using a physical scale model will provide useful adjustments. Scale model tests conducted by CNR in 1968 for the Vallabrègues spillway showed that the overall coefficient $\mu \cdot c'$ was:
– 0.25 for a height h of 0.20 cm
– 0.27 for a height h of 0.30 cm
– 0.35 for a height h of 0.40 cm
– 0.36 for a height h of 0.50 cm.

Weir submergence is an issue for front spillways but is not necessarily the same for the upstream and downstream ends of the spillway. A levee spillway, especially when it is long, is generally not horizontal, but rather sloped with an upstream-downstream slope similar to the waterlines in a leveed river. For low flows, the spillway will function in a free flow regime. As the flow increases, it will shift to submerged flow. More specifically, if the receiving area is closed, the water surface will be horizontal and water will start submerging the spillway from downstream and will gradually move upward. However, if the area is open, or if it is a channel, water will almost simultaneously submerge the entire spillway surface.

Spillway submergence will create complications if the receiving area is not closed. The flow will split into two legs and knowing the water level before this division will not be enough to determine how the flow will be distributed. A distribution hypothesis should be made to calculate the hydraulic heads based on the diffluence between both legs and then adjust the distribution to ensure both results match.

It should also be noted that proper operation of a spillway requires full understanding of the law $Q_{spillway}(H)$. Floating debris (wood, ice) and sediment can affect this law. For example, floating debris may diminish the diverted flow. On the contrary, sediment inflows may encourage diversion and trigger it for lower flood levels than planned. We will cover these topics in Chapter 4, also indicating that the spillway can be responsible for sediment deposits.

Reversing a spillway's operation

This section will not yet address spillways that are specifically designed to operate "in reverse", that is from the flooded area towards the watercourse.

When a spillway conveys water into a closed area, namely a flood expansion area or a protected area, a very strong flood might still fill the area higher than the spillway level. This mostly occurs in flood expansion areas, and we will see that this reverse operation does not necessarily mean that the flood expansion area did not attenuate the flood. This could also occur in protected areas, but only in extreme events. In that case, when the flood recedes, the area may also drain water through the spillway which would then serve as a return spillway. The submerged conditions, as described above in Section 3.2. apply to both flow directions. This explains the various cases shown in Figure 3.7.

If the spillway is very long, it may discharge water into the valley from its upstream section while also discharging from the valley into the downstream part of the river. This can be seen in the precise modelling of the Loire River Valley conducted by Hydratec at the request of the Plan Loire Grandeur Nature's multidisciplinary team of experts. This modelling was done for the longest Comoy spillways. A similar observation can be made for the Pitot spillways on the Vidourle River. The spillways farthest upstream function in submerged flow, those in the middle are drowned by high floods, and the spillways farthest downstream have "reverse" overflows. This type of operation is generally not optimal and shows how important it is to perform accurate modelling (see Section 3.7), possibly on a scale model.

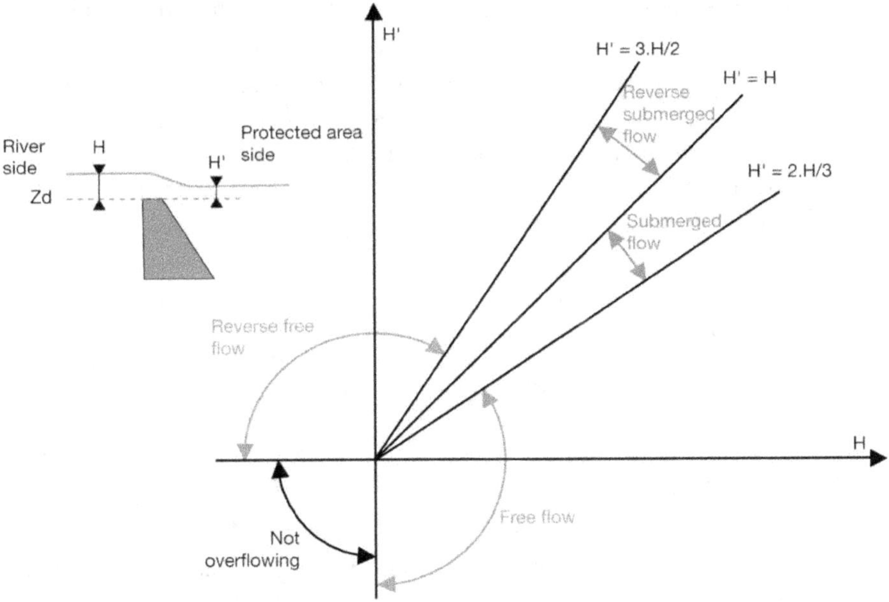

Figure 3.7. Possible flows on a spillway.

A spillway is usually designed to operate from the river to the discharge area. In will flow laterally in this direction according to the flow law described above.

When the same weir operates in the opposite direction (towards the river), it has more frontal flow. It will be either in a submerged or free flow regime according to the flow laws described in Section 3.2 with moderate flow coefficients (around 0.35) since the weir is not shaped for this flow direction. If the receiving area is closed, the waterline will be horizontal. Since the weir is usually not horizontal, the head will vary. This should be considered when calculating the discharge restored to the river.

Calculations of the return flow rate from a spillway to the river will not be used to design the structure, but to determine the shape of the hydrograph's descent and the time needed for it to go down to the spillway level. The remaining water will be removed by the drainage works (siphon, pump, check valve pipe, manual gate, etc.).

The spillway's flow law

A return spillway is specifically designed to release water from an area filled by a spillway. The diversion spillway is upstream of the area while the return spillway is downstream (Figure 1.17). Water that enters the area from the diversion spillway will flow downstream. The levee will create a head and the return spillway will prevent the levee from discharging water back into the river. This will occur when it has the same capacity as the spillway minus the hydrograph flattening effect in the flooded area.

A return spillway typically has frontal flow. The crest may be horizontal. It will operate either in submerged or free flow conditions depending on the river's water level. The flow law is the same as described above, although the flow coefficient may be higher than a spillway with a return flow. It will more likely be 0.45 to 0.5, based on the selected profile.

It is important to remember that a return spillway is not designed and does not operate as a diversion spillway in reverse (see previous section). The return spillway has a crest that is typically horizontal and shaped for flow towards the river. It will be located downstream of a long section with an upstream diversion spillway.

Spillway location

The location of the spillway on a levee that delineates the flood expansion area or the protected area may depend on geomorphological data, land use conditions at the outlet, or hydraulic conditions.

Morphological criteria

Geomorphological conditions may prevent us from placing a spillway on a convex section of the watercourse to keep sediment deposits from hindering the spillway's operation. This leaves the straight and concave sections.

The advantage of concave sections is that they strongly limit the diversion of sediment deposits towards the diversion structure thanks to the helical current that forms in the bends and the centrifugal force, which also explains the difference of water elevation in the cross-section. On rivers with high sediment transport, placement on this bank should be the priority, but this will require stronger bank protections than on straight sections to prevent circumvention. (Figure 3.8).

Installing a spillway where previous breaches have occurred could be worthwhile. Local communities are more likely to be supportive, and the levee will be protected against overflow in its weakest section. As we saw in Chapter 2 on the history of spillways, it was common for a spillway to be placed on a former breach. This also implies selection of a concave section.

The difference in water elevation in the cross-section may be significant when there is a sharp curve, and this should be considered when calculating the spillway's conveyance.

However, installing a spillway on a straight section is still a viable option, because this might be the only available option for a given site. This is particularly true when two spillways must be built across from each other.

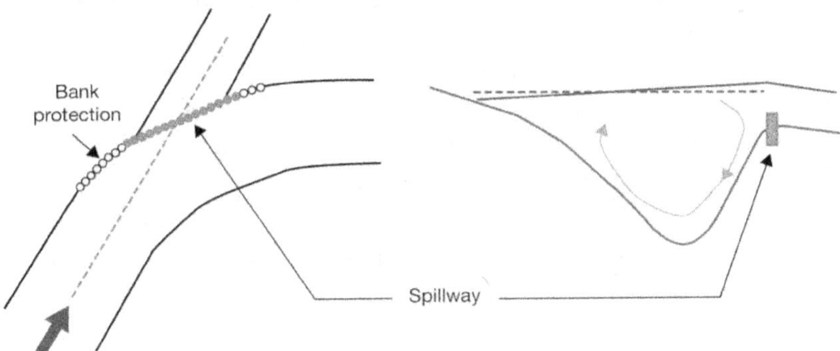

Figure 3.8. How to install a spillway in a bend.

Criteria relating to stakes

For obvious reasons, it is best to avoid building a spillway in front of important stakes that would suffer from poorly cushioned discharge. This is particularly important in protected areas. This requirement may prove very restrictive in densely populated protected areas where installing a very long spillway might be difficult. Consequently, several medium-length spillways could be built, or a single spillway with high conveyance per metre (movable, fuse plug, or labyrinth spillways, which we will describe in Chapter 5).

Hydraulic criteria

This is not a simple topic: we must examine the river's flow into the spillway, the flow beyond the spillway into the receiving area, and the flow into the river, which is affected by the diverted flow.

First, we must ensure that the structure's location will allow the river to flow easily towards the spillway, even if vegetation appears in the space between the watercourse and the levee. The hydraulic conditions mean that the water nappe in this area is necessarily low and therefore highly affected by the surface state. Furthermore, the suspended sediment load in the leveed area remains trapped between both levees. When the floodwater recedes, some of these sediments will be deposited on the unprotected floodplain that gradually increases in height (as we will discuss in Chapter 4, Section 4.3). This means we should prioritise locations that allow us to build the levee close to the riverbed.

In the river, the spillway will mainly lower the waterline downstream. It is therefore preferable to install the spillway on the upstream end to fully benefit the levee in the protected area or flood expansion area.

On the other hand, when the spillway goes into operation, it will create a stilling water cushion. As a result, it is best to choose a location that would allow for the formation of this cushion all along the levee toe before the levee is overflowed. If the receiving area has a downward slope, placing the spillway downstream might not provide enough of a water cushion upstream. For example, on the Loire River, which has a 40 cm/km slope, Comoy built spillways upstream. However, if a sufficient water cushion can be formed regardless of the spillway location, downstream placement would allow for the storage cell to be filled gradually, avoiding erosive or dangerous water velocity.

If the protected area comprises several internal storage cells separated by structuring lines, specific spillways should be built to fill each of the storage cells along the levee. But even without separate storage cells, it could be worthwhile to build several spillways in very long protected areas. Comoy initially considered this idea for several valleys, according to the narrowness of the riverbed in certain areas, for instance. This idea never came to fruition, but simulations performed in the 2000s showed that building two spillways could be useful for loading some of the valleys.

Figure 3.9 shows actual plans for a levee project in a protected area where a downstream spillway does not create a water cushion upstream when the spillway starts overflowing. An upstream installation would seem preferable here, though it only offers partial improvement in this example. Based on the topography of the receiving area, two spillways might provide a better solution to ensure the entire protected area is filled.

Lastly, in flood expansion areas (or protected areas with strong attenuation), hydraulic conditions will require choosing a location that would allow for the greatest volume of water storage to maximise flood attenuation. Though this may be achieved in any position, a downstream spillway would avoid erosive or dangerous velocities. However, if the area is long in the upstream-downstream direction, installing a spillway upstream would fill a larger part of the flood expansion area whereas a downstream spillway might leave the upstream part of the flood expansion area unused.

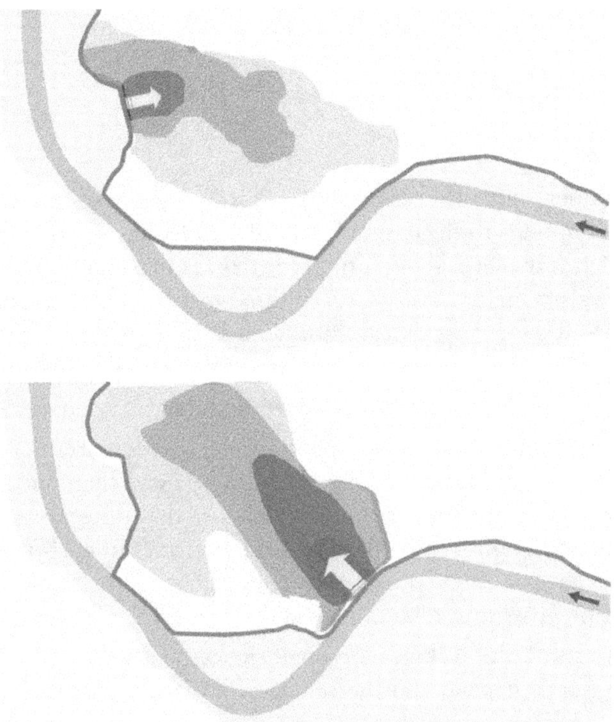

Figure 3.9. Example of a protected area. Top: a downstream spillway prevents the creation of a water cushion at the levee toe in two-thirds of the area upstream of the levee. Bottom: an upstream spillway floods a larger section of the levee toe before overflowing. For the water cushion to cover a larger area, another spillway should be built midway or water should be allowed to inflow through reshaped grounds. (Source: LRPC Blois)

Should a spillway be placed upstream or downstream?

In conclusion, we cannot dictate a definitive spillway location, though upstream installations are most frequent.

Sometimes, practical considerations (urbanisation, the river's meanders, the distance between the levee and the riverbed, actual water conveyance) will limit the options available.

Building a spillway upstream does offer the advantage of filling the entire area. If it is a protected area, this will encourage the development of a water cushion over a greater length of the levee. In a flood expansion area, this will optimise the fill volume and maximise flood attenuation. If there is a large volume of diverted flow, an upstream installation will also more effectively lower the waterline in front of the area that the spillway fills with water.

The advantage of a downstream installation is that the area will gradually fill with water without erosive or dangerous velocities. This location would seem most useful for a flat protected area without a significant flood attenuation effect.

It goes without saying that simulations should be done to help choose between several possible locations. A 2D model would be necessary in this case (see below).

Protection, safety, and danger floods

Concepts, definitions

A review of dams

As a reminder, dams have a maximum water level (also called the safety level), which corresponds to the design flood (also called the safety flood), and a danger level, which is the level above which the dam may suffer major damage that may quickly lead to failure. For an embankment dam, the danger level is generally the same as for the crest. For flood-attenuating dams with open sluices, there is also a protection level defined as the surface flood spillway level (see Figure 3.10).

The danger level is reached when a so-called danger flood enters the reservoir.

Each flood-attenuating dam has a protection level that corresponds to a protection flood entering the reservoir. There may also be more than one protection level if several hydrograph shapes are considered. For the protection flood, flow that is released downstream is determined by the bottom outlet. If the dam is properly designed, this flow corresponds to the level that is acceptable in the areas to be protected downstream, taking intermediate inflows into account.

Danger and protection floods are intrinsic to the specific dam. The design flood (or safety flood) depends on the results of a hydrologic study and the choice of flood frequency.

We will apply these concepts to levees, although this poses a challenge since the levee, waterline, and spillway profiles in this case are not horizontal, nor even necessarily parallel.

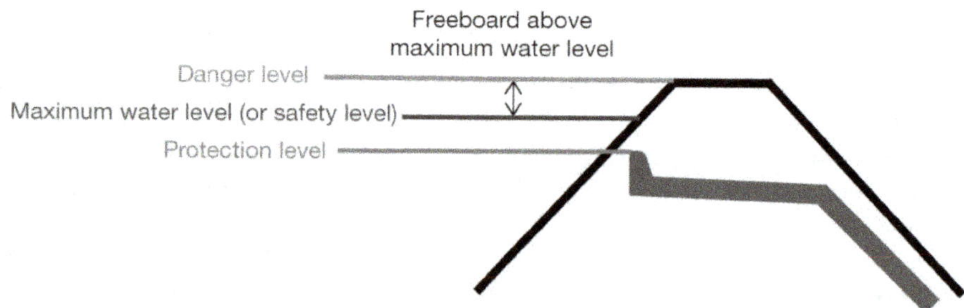

Figure 3.10. Danger level, maximum water level, and protection level for a flood-attenuating earthen dam.

We will use the word "level" to refer to the waterline, flow, or degree at a specific location and not to the topographic elevation of a given point. For instance, a flood scale: danger level, safety level, protection level.

Can we define the levee's danger level?

Like earthen dams, an earthen levee's danger level typically refers to the floodwaters that near its crest. If the levee has an uneven longitudinal profile, the danger level has been attained when floodwaters reach at least one of the low spots, as shown in Figure 3.11. However, if the levee has been designed so that these low spots can resist overflows, they can be ignored. This is how the Vidourle River Union adapted its levees without changing their profiles and by "reinforcing" some low spots. On the Vidourle River downstream from the city of Lunel, the reinforcement of a 20 cm low spot increased the flow rate of the danger flood from 860 to 930 m^3/s (ICAT, 2006).

However, the waterline is not necessarily parallel to the levee crest, especially if it has a spillway.

Let's look at a levee with a spillway in a protected area or a flood expansion area. For a typical subcritical flow rate, the same upstream flow $Q_2=Q_1$ will result in a waterline L_2 that is lower than L_1 both upstream and downstream of the spillway, for two different reasons as previously described (Figure 3.11). However, the waterline will not be lowered far enough upstream of the spillway. Therefore, the danger level remains unchanged far upstream: $Q_2=Q_1$. In contrast, L_2 maintains a freeboard downstream. A $Q_3> Q_1$ flow should be introduced upstream so that the waterline L_3 can reach the crest of levee. The danger level downstream will then be $Q_3> Q_1$.

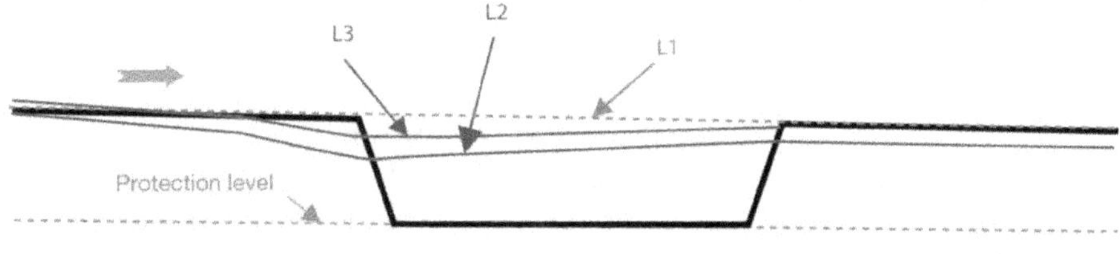

Figure 3.11. Change in the waterline for a danger flood after adding a spillway:
- L_1 is the waterline for danger flood Q_1 without a spillway.
- L_2 is the waterline for the same flood with a spillway (the same flow $Q_2 = Q_1$ upstream, the waterline is lowered).
- L_3 is the waterline that reaches the crest of the levee downstream of the spillway (upstream flow $Q_3> Q_1$).

We can see that it is impossible to define a single danger level for a levee. We must consider the entire levee system in terms of overflows for either a protected area or a flood expansion area.

A danger flood that reaches the danger level in a flood expansion area or protected area might cause levee failure in at least one spot in the area under consideration.

For earthen levees, a danger flood is the weakest flood that causes overflow out of the spillway if the levee remains structurally intact up to this point. If a spillway is built upstream of the flood expansion area or protected area, it increases the flow rate of the danger flood by the value of the diverted flow. This is obviously a good thing. If the spillway is built all the way downstream of a long flood expansion area or a protected area, the flood at danger level will not change in comparison to the same levee without a spillway.

The design of the danger level of an earthen levee should be carefully reviewed on a case-by-case basis. The baseline will be a calculation of the waterline, but everything that impacts the actual water level must be taken into account:
– Raising of the water level in the extrados of a bend (except if the calculation is based on a 2D model that already includes this effect);
– Potential backwater effect downstream due to a predictable jam at a bridge, a weir, or a sharp bend;
– Role of sediment deposits via bedload, or, on the contrary, incision (see Chapter 4), except when using a mobile-bed model;
– Standing waves that form due to strong discontinuities in a supercritical flow section.

Can we define a levee's safety flood?

The concept of design flood (or safety flood) for dams may be applied to levees. It is the flood level at which the levee maintains a margin of safety that protects against various sources of failure. The safety flood must:
– offer some freeboard in relation to the waterline of the danger flood;
– ensure the mechanical stability of the levee with sufficient safety coefficients, midway between the coefficients obtained in a normal situation corresponding to the average watercourse level and the coefficients required for the extreme case of danger flood;
– resist internal erosion;
– ensure the spillway functions properly, with no damage.

In France, the freeboard calculation is based on waves to prevent excessively frequent overtopping of the levee crest. Waves generated by wind can measure several decimetres when there is a great distance between levees or when long straight sections extend in the direction of the wind.

For a new levee project, the designer will define a safety flood with a specific return period. The relevant waterline will be calculated and the waterline of the danger flood will then be determined to comply with the calculated or estimated freeboard value all along the levee. This is how the longitudinal profile is calculated for the earthen levee.

For an existing levee, we use the same method in reverse. We start with the danger flood, where the floodwater comes close to the lowest points of the levee; then we determine the safety flood to comply with the calculated or estimated freeboard value all along the levee; then we determine the return period for the safety flood and compare it to what is considered ideal.

As with the danger flood, the safety flood will be determined for the entire levee system, protected area, or flood expansion area.

The safety flood for a flood expansion area or protected area is designed to preserve a margin of safety against various sources of failure in all parts of the area that need protecting (freeboard for waves and the safety coefficients needed to ensure stability).

Protection flood for the area

The protection flood in a flood expansion area that fills with water through a spillway or a protected area that is protected by a spillway will be the lowest flood that starts overflowing the spillway. This implies that the levee has been able to resist until then. Here we are not referring to the levee's protection flood: if the levee is properly designed, built, and maintained, it should resist much higher floods.

In a protected area, the protection flood will be the maximum flood that remains acceptable in the locations requiring protection. When floodwaters reach above the protection level, the spillway goes into effect and the areas that were initially protected will gradually be flooded. However, if there is no human presence in the low parts of the protected area, the acceptable flood may be higher than the protection flood (all the better).

A flood expansion area works a bit differently. In this case, the design will focus on the acceptable discharge in areas farther downstream. If this discharge is used to define the protection flood, any floods that reach above that discharge will be split between the watercourse and the flood expansion area. But the acceptable discharge downstream will necessarily be exceeded. Therefore, the discharge of the protection flood in the flood expansion area must be lower than the acceptable discharge downstream (taking any intermediate inflows into account). Otherwise, floods that could damage the downstream area would be allowed through without attenuation, and attenuation would only begin with very damaging floods (we will return to this topic later in the chapter).

The case of a levee without a spillway

This section only applies to protected areas since flood expansion areas cannot function without a spillway. There may be siphons or pumps, but they can be considered spillways for our purposes (the spillway level will then be regarded as the priming level).

Let us examine a protected area without a spillway (Figure 3.12). The danger flood (Q_1) will create a waterline (L_1) that approaches the levee's low spots along the crest – assuming the levee remains structurally sound. The standard is for the protection flood to coincide with the safety flood. It corresponds to a waterline that leaves some freeboard in relation to the levee crest. Since there is no indication of the protection flood, nothing will warn us it is being exceeded.

The advantage of this standard is that in both cases, the following order occurs:

Protection flood ≤ safety flood < danger flood.

When comparing a levee in a protected area without a spillway with the same levee with a spillway, the spillway obviously lowers the protection flood level. However, it creates a beneficial gap between the protection flood and the safety flood. There is some leeway before the levee's resistance is no longer guaranteed.

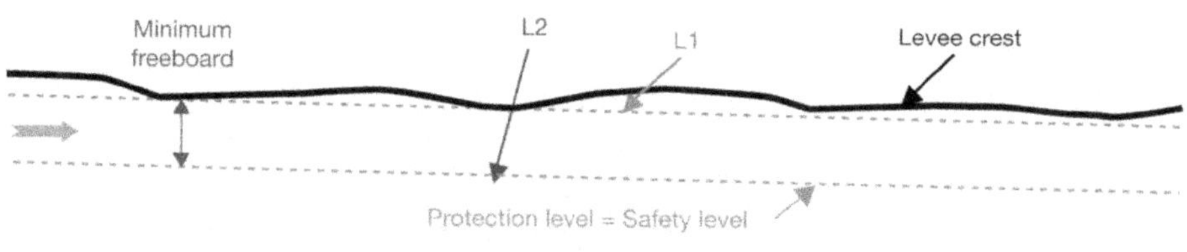

Figure 3.12. Waterlines for danger and safety levels when there is no spillway:
- L_1 is the waterline for the danger level.
- L_2 is the waterline for the safety level that maintains a freeboard before the crest is reached.

In summary

• **The danger flood** in a levee system is the one at which levee failure is possible in at least one spot in the area being considered; if the (earthen) levee remains structurally sound, it will be a flood in which the waterline comes close to a low spot on the crest; but if the levee has not been properly designed, built, or maintained, the danger flood will be lower and harder to assess; when this flood level is reached, the probability of levee failure is high, e.g., about 50% (Figure 3.13.).

• **The safety flood** will help maintain a margin of safety in comparison to the danger flood, both in terms of hydraulics and stability; when this flood level is reached, the probability of levee failure will be low, e.g., about 1%.

• **The protection flood** is the point at which water starts overflowing the spillway; when this flood level has been reached, the probability of levee failure is almost non-existent (providing the levee remains structurally sound).

• When there is no spillway, the protection flood is considered equivalent to the safety flood and cannot be distinguished.

• The protection flood is typically the same as the acceptable flood in a protected area; above this level, the protected area will start flooding.

• For spillway conveyance purposes, the protection flood must be lower than the acceptable downstream flood level in a flood expansion area, but not by much.

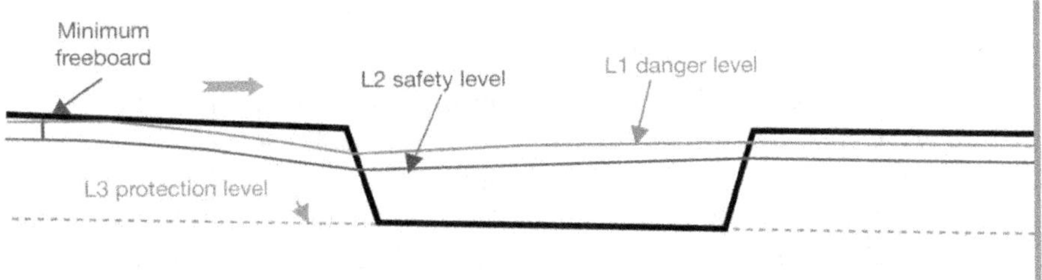

Figure 3.13. Waterlines for danger, safety, and protection floods in a protected area or flood expansion area:
- L_1 is the waterline for the danger flood (non-existent freeboard on at least one spot in the protected area or flood expansion area).
- L_2 is the waterline for the safety flood and maintains a freeboard all along the levee before the crest is reached.
- L_3 is the waterline for the protection flood (it overflows on at least one part of the spillway).

As an example, here are the key figures for a planned protection levee designed by the Syndicat du Vidourle in Gallargues (for a secondary levee):
– Danger flood discharge = 3,000 m³/s (1,000-year flood with peak flow adjustment)
– Protection flood discharge = 2,400 m³/s (flood of 9 September 2002, with an estimated 400-year return period)
– Spillway dimensions: 30 m long, 60 cm drawdown
– Spillway flow for the danger flood: 14 m³/s

This shows that the spillway's capacity is very low relative to the floods in the watercourse, meaning that flood attenuation is clearly not the intent in this example.

Considerations for human and levee safety

Note: we use the terms "safety" and "security" synonymously, as is the case in many regulatory documents, etymologically, and in informal language. It might be preferable to use the term "safety of a structure" and "human security," but this is not common.

The government establishes guidelines for the safety of hydraulic works and is responsible for ensuring compliance since human security is at stake.

Managers of hydraulic works should have the same safety objectives since they will be held liable if their structure is responsible for damage to individuals (or their property). Even in the absence of damage to third parties, managers should prevent any damage for financial reasons.

Humans may be threatened:
– by a structural failure in the hydraulic work (e.g., a gravity structure is overturned, internal erosion or overflow causes a breach in an earthen levee or dam) that suddenly releases a flood wave;
– by a functional failure in the hydraulic work or something appurtenant to the work (e.g., the deliberate or accidental opening of a dam gate drowns people

in the downstream valley); this is less likely with a levee, but it could occur if a movable weir system suddenly collapses;

– by a hydraulic work that is overtopped but not damaged. In this case, it is the water flowing over the dam or levee that will cause damage, such as water spilling over a coastal or river levee without breaking it, or a high wave flowing over a dam due to sliding in the reservoir. In the first case, it is the effect of the waves that caused the damage, not the structure. In the second case, the structure is to blame since the reservoir caused the wave to form.

The structure's safety and human safety are two concepts that cannot be dissociated, yet they should not be confused:

– A structure may cause casualties without failing, as with the Vajont dam or an ill-timed dam gate operation.

– On the contrary, a structure may fail without causing casualties. In France, many small dams or levees have failed without any casualties, or even without any damage beyond the structure itself.

We can also see that the concept of human safety does not just depend on controlling the water level upstream of the structure. The water level is clearly to blame when resistant works are overtopped or when they fail because of overflow. However, this is not the case for an ill-timed bottom gate operation since the flow rate has little connection to the water level in the reservoir.

Let us now examine the case that interests us: earthen river levees.

If the levee has a spillway in good condition, the levee's safety level will be higher than the protection level. However, above a particular flow rate, even a spillway that functions properly can endanger people that have not been moved to safety. In this case, we have the following order:

Protection level < human safety level ≤ levee safety level

If the levee does not have a spillway, there is no active or passive device to bring water into the protected area. Water only enters if there is overflow without levee failure, or because of levee failure (caused by overflow or before overflow).

In the first case, with an overflow-resistant levee, we have the same order as before (as if the levee were a large spillway):

Protection level < human safety level ≤ levee safety level

In the second case, which is quite common, humans are threatened at the exact moment of levee failure. The order is then:

Human safety level = levee safety level

In this same case, the actual protection level is the same as the safety level (by definition): the apparent protection level may be higher and therefore give a misleading impression of safety.

In the end, the human safety level is the same as the safety level of the structure for levees that do not resist overflows and do not have a spillway, as is often the case. Otherwise, the human safety level will be lower or equal to the safety level of the structure.

> **In short, the order will always be:**
>
> Human safety level behind the levee ≤ levee safety level.

Any difference between these two levels is a good thing, as the risk due to levee failure will inevitably be higher than when there is no failure. The difference between both levels expressed as the waterline will often be small. In reality, spillway drawdowns are usually just a few decimetres. As a result, the safety level may only be 20–80 cm higher than the protection level. Both human and levee safety levels are determined by calculations or measurements with margins of error, and the designer will often select identical values – which is not necessarily a problem.

The principle of flood attenuation or flood hydrograph flattening

Natural flood hydrograph flattening

Natural flood expansion areas can serve to flatten flood hydrographs. This means that the downstream flood hydrograph is less pointed than the upstream hydrograph (Figure 3.14). The stored volume (dark blue) and the discharged volume (light blue) are identical if infiltration is negligible, which is generally the case. Since the downstream peak discharge is lower, the maximum water levels are lower downstream. They are also lower to a certain extent upstream due to the backwater effect (for subcritical flows). The benefit of hydrograph flattening is cumulative along the watercourse.

The same flattening effect will occur in a dam reservoir that is full before floodwaters pour in: the peak outflow (through the spillway) will be lower than the peak inflow.

In this handbook, we use the term "hydrograph flattening" rather than "flood attenuation": the former means that the hydrography is "rolled out", as in a rolling mill, while maintaining the same volume; the latter means that the hydrograph peak is removed. In a natural flood expansion area and a dam with a full reservoir, the arriving flood cannot be attenuated; its hydrograph can only be flattened. However, with a flood-attenuating dam (or in a flood expansion area), both flood attenuation and hydrograph flattening will occur simultaneously.

This handbook will differentiate between these two ideas: hydrograph flattening maintains the volume of the water (Figure 3.14), while flood attenuation does not (or does so much later).

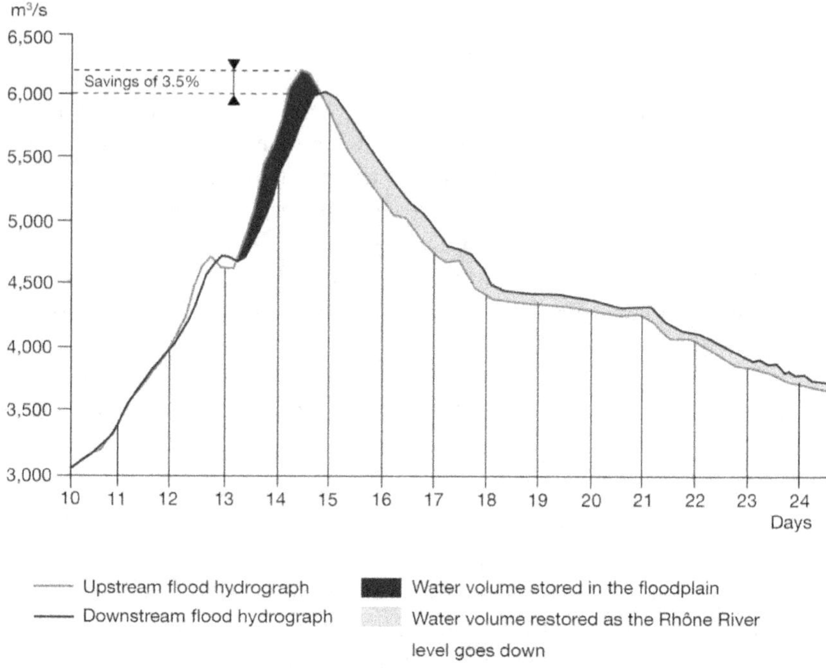

Figure 3.14. Example of flood hydrograph flattening in the Tricastin plain for a medium flood on the Rhône River (about 100-year level). (Source: Overall Study of the Rhône)

The principle of organised flood attenuation

We will now examine flood expansion areas where spillways divert a volume of water. The intent is to divert this volume at the right time to reduce downstream overflows in areas with important stakes (either other levees or inhabited areas without levees).

Figure 3.15 shows the flood attenuation principle in a flood expansion area that is empty before flooding.

At time t_1, the watercourse flow is higher than flow Q_p at the protection level. The spillway will begin operation. As the river level gradually rises, the flow diverted towards the flood expansion area will increase. At time t_2, when the natural flood has reached its peak, the diverted flow will also be at its peak. At time t_3, when the natural flood has gone back down to Q_p, the diverted flow ceases.

Let us look at the example of a flood expansion area with an outlet pipe that opens onto the riverbed with a check valve. At time t_4, the river's flow is under bankfull discharge Q_b and the flood expansion area drainage device goes into operation and gradually empties the area. However, we will not take this drainage step into account here since it has little impact on the spillway design.

In the lower diagram in Figure 3.15, the water level in the flood expansion area increases starting at time t_1, and will keep on increasing after time t_2, though less quickly (inflexion point). The water level will reach its peak at time t_3, and in this example, the level in the flood expansion area will remain lower than the spillway level *(zs)*. The flow remains free and is not conveyed from the flood expansion area towards the river. We will examine later what happens with flood attenuation when the flood expansion area level goes above the spillway level.

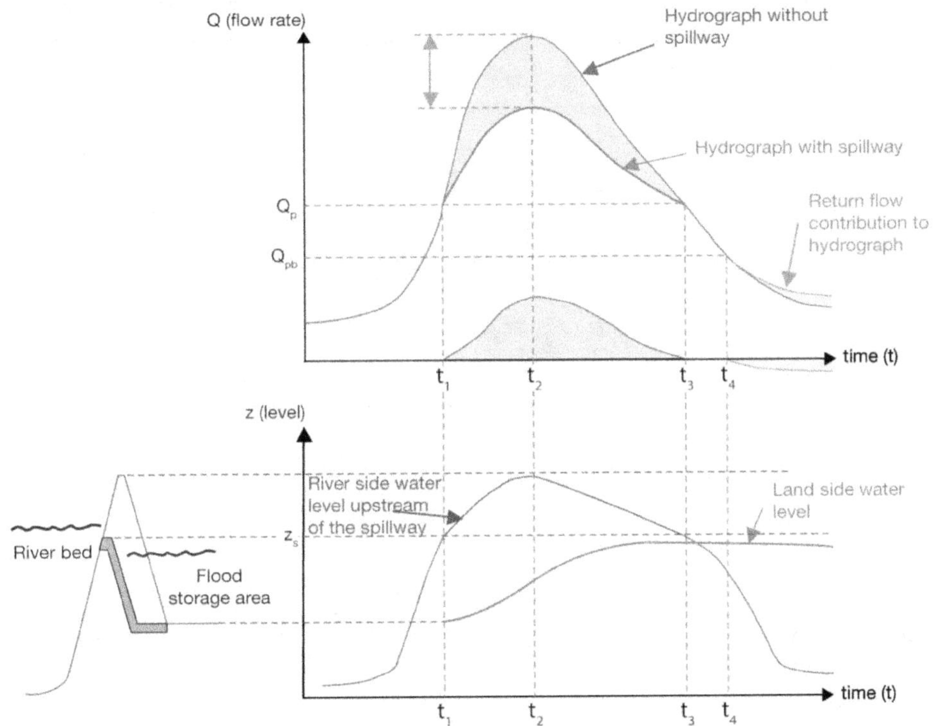

Figure 3.15. Principle of flood attenuation by discharging water into a flood expansion area. Blue: volume diverted towards the flood expansion area. Yellow: (identical) volume restored after flooding. Bottom: flood expansion area water level (compared with the spillway level). Note: there is no unprotected area in the small levee diagram.

General hydraulic design principles

The hydraulic design of a spillway mainly consists in calculating its level (or more specifically the longitudinal profile of its crest) and overflow length. We will first consider flood expansion areas, followed by protected areas. The hydraulic principles are obviously the same for both, so we will not address them again for protected areas. However, decisions must be made based on fundamentally different criteria.

We will illustrate our analysis with numerical examples. Numerical data, lengths, flows and elevation levels are not essential but make the comments easier to read. Our calculations are based on a flat-bottomed flood expansion area without a delayed effect, as in a bathtub.

Determining the length of a spillway in a flood expansion area

We have set a fixed level for the spillway and simply varied its length. For a short spillway, the selected flood will not fill the entire flood expansion area, which remains below the spillway level. In the example in Figure 3.16, the peak flow is attenuated by 16%.

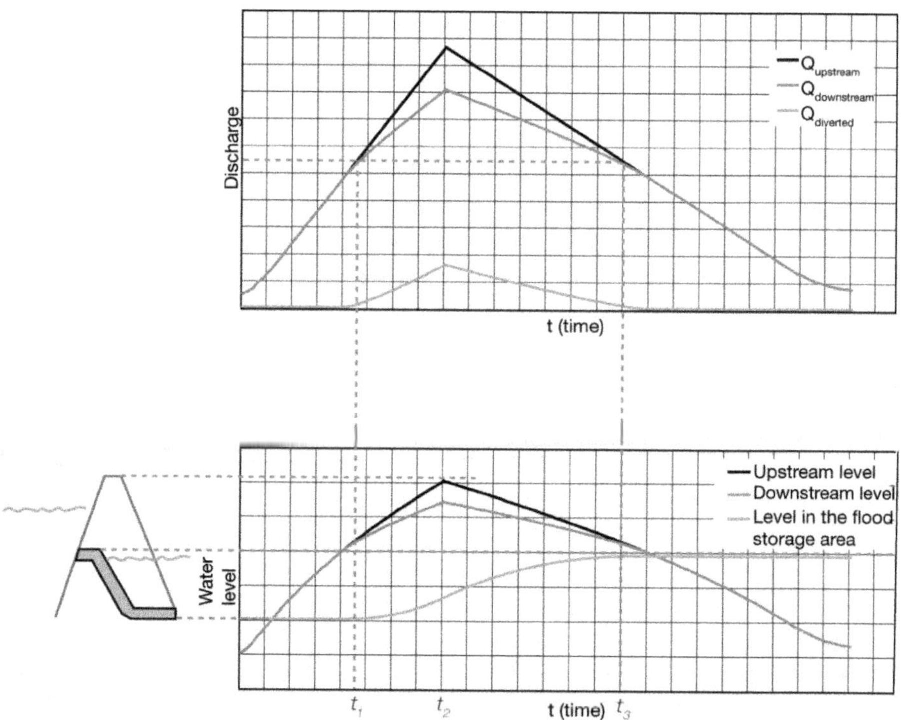

Figure 3.16. With a short spillway (L= 20 m), the flood expansion area is not entirely filled.

We then model the same flood with a much longer spillway (Figure 3.17).

At time t_5, the flood expansion area will fill up to the level of the spillway but the spillway will remain in free flow. At time t_6, the flood expansion area level will be high enough for the spillway to start being drowned at the downstream end; the flow will be lower than in free flow conditions (dotted line). The flow will be cancelled out at time t_7 when the river and flood expansion area levels reach an equilibrium. The flow direction will therefore be reversed and the flood expansion area will drain towards the river until the weir level is reached (at t_9).

This does not affect the efficacy of the flood expansion area: a second flood peak starts to appear, though lower than the first attenuated peak (Figure 3.17). This time, the peak flow is attenuated by 26%. The second peak is not harmful and this spillway more effectively attenuates the flood.

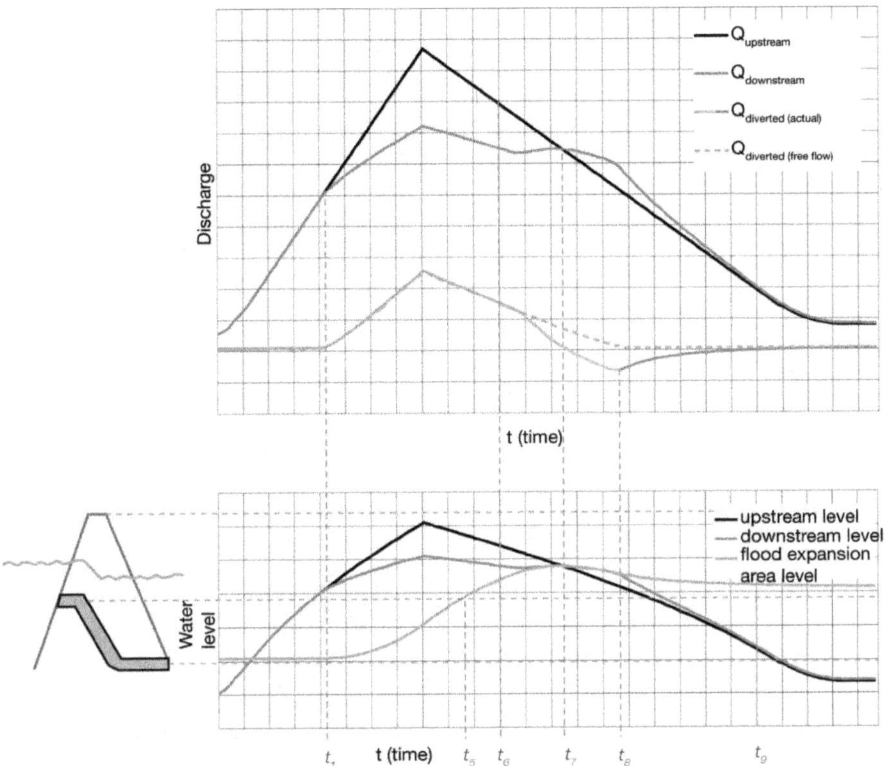

Figure 3.17. Beginning of flood expansion area engorgement with a longer spillway (L=50 m). The dotted-line hydrographs would occur if the filling of the flood expansion area were not a limiting factor. Flow towards the flood expansion area is free until t_6 (blue line) then submerged until t_7 (green). It then reverses, with a submerged flow until t_8 (orange), and later free flow (red).

We will now make the spillway even longer (Figure 3.18). This time, the second peak at time t_7 will be higher than the first peak. Consequently, the peak will only be lowered by 26% whereas the inflow peak is lowered by 33%. We achieve the same flood attenuation effect as in the previous example, even though the spillway is longer. This clearly shows that this additional length is not useful, at least for this particular flood.

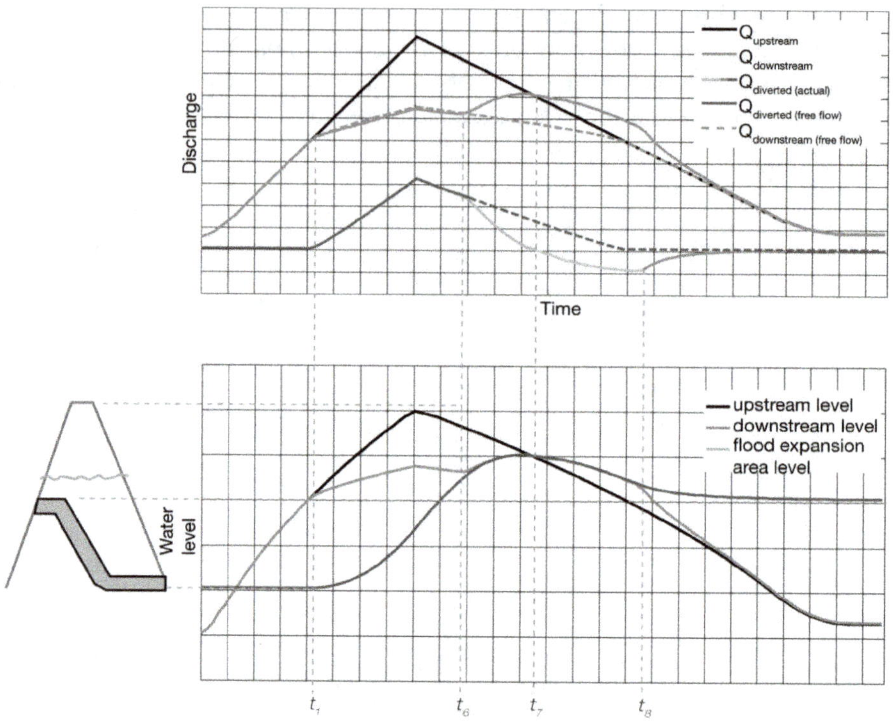

Figure 3.18. Beginning of flood expansion area engorgement with a very long spillway (L=110 m). The draining of the flood expansion area will reduce the benefit of flood attenuation. The dotted-line hydrographs would occur if the filling of the flood expansion area were not a limiting factor. Flow towards the flood expansion area is free until t6 (blue line) then submerged until t7 (green). It then reverses into submerged flow until t8 (orange) and later becomes free flow (red).

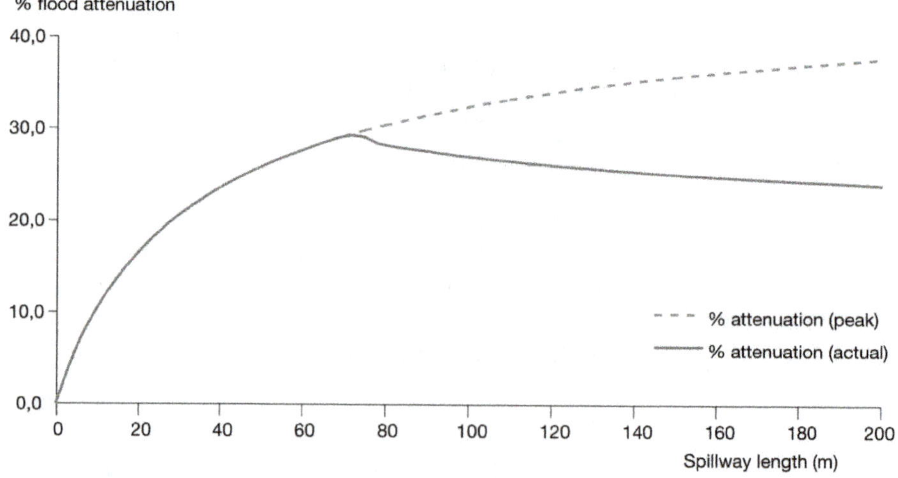

Figure 3.19. Optimum attenuation for a given flood with a given crest level. The dotted lines show the flood attenuation effect for a very large flood expansion area or if water is discharged into a flow channel that is not drowned.

We simulated many different spillway lengths to obtain the diagram in Figure 3.19. It shows the spillway length that will produce the best attenuation for the flood in question. Flood attenuation of more than 30% is impossible. To obtain the maximum level of attenuation, the spillway should be 72 m long.

However, various flood levels must be considered. To do so, we decided to vary peak flows while maintaining the shape and duration of the hydrograph. This means that floods are inferred from one another by affinity on the flow axis. Figure 3.20 shows the result. A 46 m spillway provides optimum attenuation for a 270 m^3/s flood (blue line). A shorter spillway offers better attenuation for stronger floods and vice versa. Discharge below 122 m^3/s will not enter the flood expansion area (protection flow rate of the area).

For example, a 46 m spillway will offer optimum attenuation for the 270 m^3/s flood (28%). Flood attenuation is still provided for a strong 300 m^3/s flood (25%, close to the 26% optimum). Flood attenuation will also occur for a medium 240 m^3/s flood (25%, with 29% as the optimum) and a moderate 200 m^3/s flood (18%, with a 22% optimum). If the acceptable downstream flow is Q_{acc} = 160 m^3/s, considering intermediate inflows, the spillway makes all peak discharge below 200 m^3/s acceptable. This choice of spillway length seems satisfactory, and could even extend to 70 m but at greater cost. With an even longer spillway, medium floods would be prioritised at the expense of strong floods. The final choice should therefore be based on a cost-benefit analysis.

We can also see that, unlike a flood-attenuating dam, it is impossible to attenuate all the extra flow above the harmless flow level. The floodwater will be split between the flood expansion area and the leveed riverbed. The flood expansion area limits the increase of the water level, but cannot stop it.

Determining the spillway level in a flood expansion area

We will now keep the same spillway length but vary its level. We use the term "level" to simplify things. We should actually refer to the longitudinal profile of the crest since it typically runs parallel to the waterlines during flooding. We will continue to use the same flood expansion area in all our examples.

Figure 3.21 shows that a spillway that is too low (3.5 m) will only attenuate small, harmless floods and would not be very effective for strong floods. It is important to understand that a flood expansion area should not be filled with water too early and too quickly so it will still be effective when the peak flood flows through.

However, this figure also shows that harmless floods should be allowed into the flood expansion area; otherwise, the first harmful floods would not place a significant load on the spillway and would therefore not really be attenuated. In this case, a 4.5 m spillway is too high since the first harmful floods will not really be attenuated. This choice is therefore untenable: it is possible to justify a spillway offering only partial improvement for a very rare event; but it is harder to explain that it is useless for a more common, yet harmful, flood.

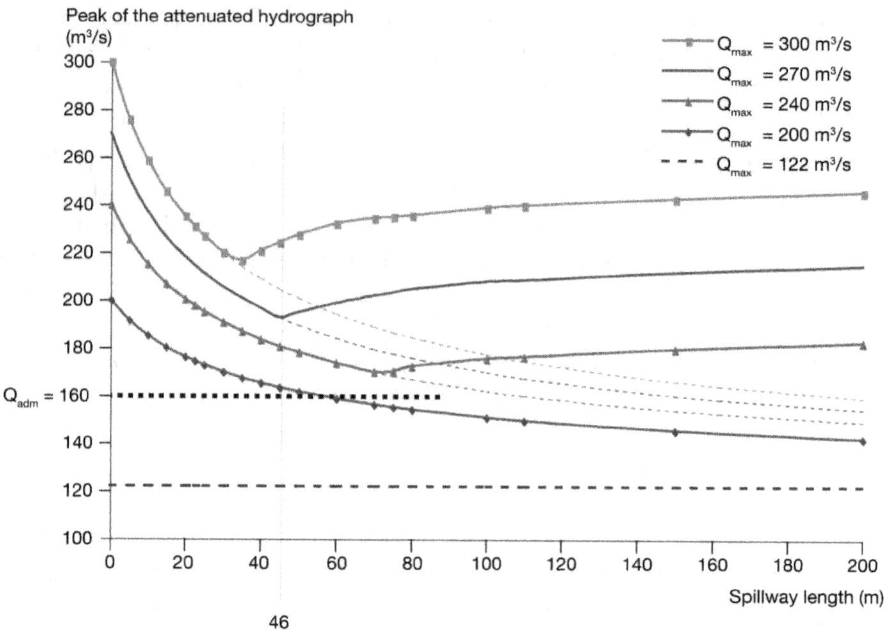

Figure 3.20. Benefit of a spillway in a flood expansion area for various upstream flow rates when changing the length of the spillway. The dotted lines show what the downstream flow would look like for a very large flood expansion area.

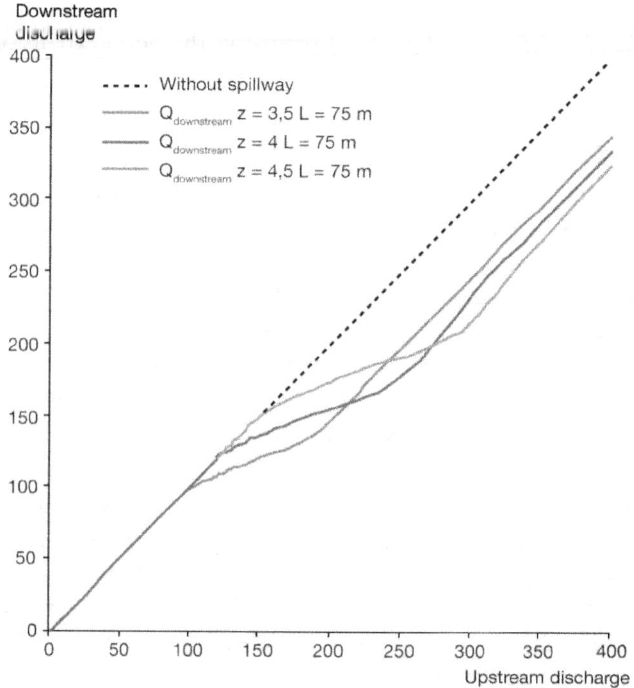

Figure 3.21. Benefit of a spillway in a flood expansion area according to the upstream discharge by varying the spillway level.

Determining the level of the spillway in a protected area

In a protected area, the level of the safety spillway is automatically determined by the value chosen for the protection flood. For this flood, calculating the waterline of the riverbed will help determine the profile of the spillway crest. In theory, the profile should come close to this waterline, although a margin of safety of several decimetres is advisable to take into account any approximations in the calculation.

Determining the spillway length in a protected area

For a spillway in a protected area, the selection criteria are different than in a flood expansion area, although the calculation remains the same. It is even simpler since we do not need to take submerged flow into account. Since the levee and spillway levels are fixed, the spillway length will vary until the volume discharged by the danger flood forms a substantial water cushion (Figure 3.22). When a higher flood causes water to overflow the earthen levee, the downstream toe will absorb the water energy. Backward erosion will be significantly delayed, especially if the overflow occurs across the entire structure with a consistent cushion. If the flood is even higher, the water cushion will thicken and improve protection at the toe. If the levee does end up failing, the water energy will be absorbed and the area will slowly fill with water. Of course, all local inhabitants must be evacuated or moved to safety, as we will discuss later (Chapter 6). Clearly, the smaller the protected area and the flatter the bottom, the easier it will be to create a sufficient water cushion. The ideal water cushion should be at least 50 cm everywhere.

If there is a slope in the protected area, the water cushion will be thicker downstream and might be insufficient or even inexistent upstream (see Figure 3.9). In this case, the longitudinal profile of the levee crest may be adjusted: water will first flow downstream towards the protected area, then gradually upstream as the water cushion thickens.

It is also possible to encourage the formation of a water cushion in certain spots using modelled grounds or longitudinal linear embankments.

Lastly, the levee toe can be protected by riprap in areas where the water cushion is too thin.

The parameter

$$\Delta t = t_{danger} - t_1$$

specifies the time available to evacuate inhabitants or bring them to safety. The lower the flood rise gradient the more time there is, but we have no control over this parameter. We can extend the time by altering the difference between the elevation of the spillway and the crest of the levee, but there is little room for manoeuvre here. We can also adjust the spillway length, but this has only limited impact on the flood hydrograph. In areas with a steep gradient increase, an early evacuation must be planned using a flood forecasting model.

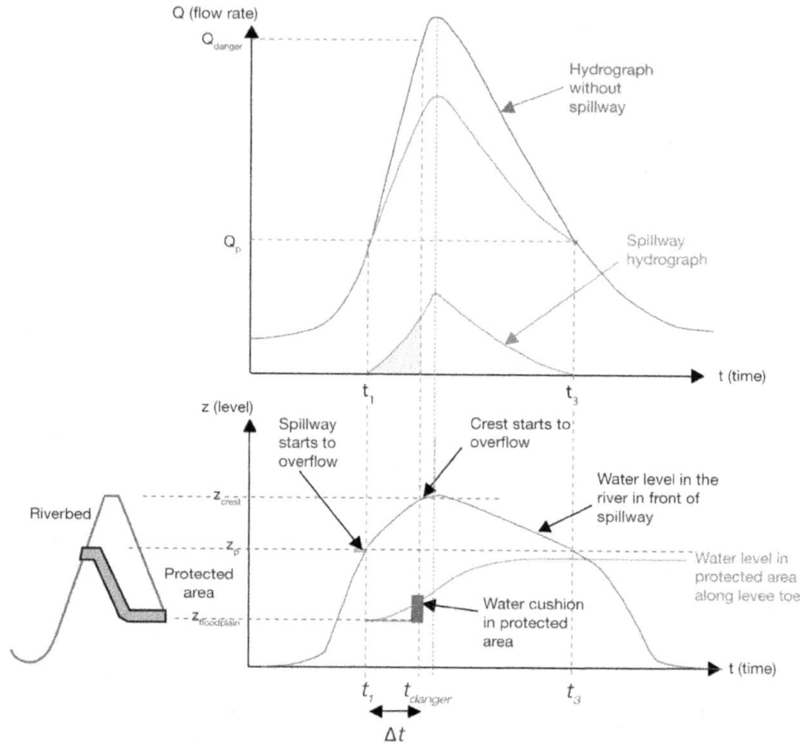

Figure 3.22. Calculating the volume discharged into a protected area. The spillway is the right size when the volume (yellow) produces a water cushion of at least 50 cm in the entire area at the levee toe. The flow discharged has been exaggerated for the purposes of the drawing.

Since we do not want to discharge too much floodwater into the protected area, it is best not to install a spillway that is larger than necessary in a protected area.

Lastly, it is important to note that the spillway reduces but does not eliminate the risk of a breach. The spillway has a positive effect since the impact will be absorbed by the water cushion at the moment the rapid flow over the face reaches the horizontal ground at the levee toe. However, the spillway will not eliminate the risk of erosion on the face, only at the base, all the more so if the levee is poorly compacted and not well covered in grass.

A spillway will reduce but not eliminate the hydraulic head between the leveed riverbed and the floodplain. Safety measures should therefore be put in place (safe floors in buildings, preventive evacuation), which we will discuss further in Chapter 6.

Stilling basin at the levee toe

The smaller and flatter the protected area, the easier is to generate the complete water cushion mentioned previously. If topography prevents the formation of a water cushion across the entire surface, another option is to create a buffer zone that can absorb all the erosive energy from the discharge flowing over the levee.

This zone can be created by excavating the entire surface or building a longitudinal embankment. To the extent possible, the spillway should fill this zone with water before overflow occurs on the levee, with about 50 cm of water at the levee toe spreading over at least 5 m in width and along the entire length of the levee.

To do so, there must be a strip of open land between the levee and the stakes in the protected area.

This entire buffer area, which is clearly meant to store water (therefore without an average flow velocity), must have an easement. This strip of land will also provide an access path for monitoring and maintenance purposes and serve as a walking path.

Energy dissipation

As water flows over the spillway chute, its energy strongly increases since it has a highly supercritical flow. Away from the toe, the flow will usually become subcritical, forming a hydraulic jump. This hydraulic jump will dissipate energy. If no measures are taken, the hydraulic jump will create an erosion pit and may even dislodge the end of the spillway chute. Experience with flood spillways on dams suggests that it would be worthwhile to include a stilling basin that is sized to contain the hydraulic jump. However, this kind of structure is rather large, and designers may, without saying so, end up building smaller structures designed to prevent the chute from being dislodged, but not to prevent erosion. We will discuss this issue later.

Hydraulic jump basin

Protection can be offered by a hydraulic jump basin (or stilling basin) that is long enough to contain the hydraulic jump and reduce the residual energy output (Figure 3.23). We will show examples of existing basins in Chapter 5 (Photo 5.1 and Figures 5.7 and 5.9).

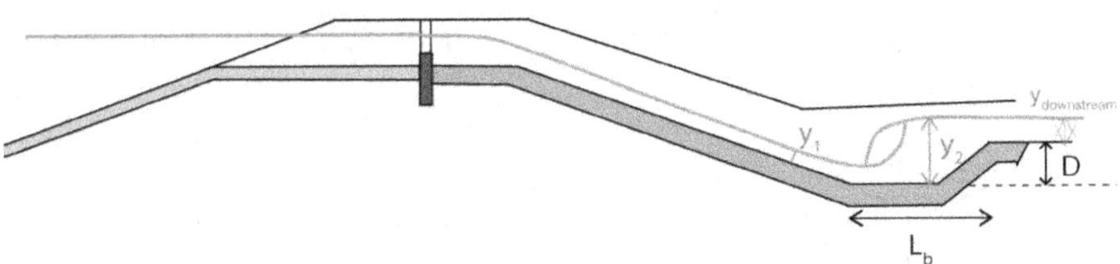

Figure 3.23. Hydraulic jump basin at the spillway toe.

We need to calculate the depth of the basin and the length of the hydraulic jump at the spillway toe.

Depth can be calculated using hydraulic jump theory. For a given flow (q) per unit of width, upstream conditions will determine flow depth (y_1) before the hydraulic jump. Using the Bernoulli equation,

$$H_s = y_1 + V_1^2 / 2g$$

with $V_1 = q / y_1$ if the channel is rectangular. H_s is the specific head at the downstream toe, which is slightly lower than the head upstream of the weir if we consider the loss of head on the downstream face.

Solving this equation (of 3rd degree) with the known flow value gives a resulting value y_1. This immediately gives us velocity

$$V_1 = q / y_1$$

and the square of the Froude number

$$F_1^2 = V_1^2 / (gy_1).$$

The hydraulic jump formula will then provide the downstream flow depth (or conjugate flow depth):

$$y_2 = (\sqrt{1 + 8F_1^2} - 1) \, y_1 \ / \ 2.$$

We then get depth D for the basin by making sure that the downstream flow depth y_2 coincides with the level $y_{downstream}$ required by downstream flow into the plain for the flow in question. The depth of the basin below the natural ground level is

$$D = y_2 - y_{downstream}.$$

This calculation should be done for various unit flows q, and typically the maximum value for D will be retained. The strongest flow will not necessarily produce the deepest setting. This is the case when the weir is flooded: there are no more supercritical flow sections or hydraulic jumps.

If the flow depth in the plain is lower than the conjugate flow depth, the hydraulic jump will become unstable and some of the energy will dissipate beyond the basin. Otherwise, the hydraulic jump will be submerged and a larger area will be required to dissipate the energy. However, this situation is still beneficial, since there is a downstream water cushion to limit the erosive power of the discharge. And if the downstream level is very high, there will be neither hydraulic jump nor extra energy to dissipate.

Estimation of the hydraulic jump length (L_r) will be based on empirical results and depend on the shape of the downstream face.

In a rectangular channel with a horizontal bottom, it can be calculated using the following approximate formula:

$$L_r = 35y_2\sqrt{F_1} / (8 + F_1),$$

which is valid above $F_1 = 3$ (Sinniger and Hager, 1989). Basically, the length of the jump is six times the conjugate flow depth with a minimum Froude number of 4. That is $L_r \simeq 6y_2$.

Lastly, if the hydraulic jump is flooded from downstream ($y_{downstream} > y_2$), the length of the submerged hydraulic jump will be higher than these values and will be

$$L_r = 4.9y_{downstream} + 1.2 \, y_2$$

according to Lencastre (1996). Indeed, $L_r \geq 6.1y_2$, which means that the hydraulic jump will need a larger area than when it is not flooded.

At any rate, the length of the basin is usually accepted as $L_b = L_r$.

The lengths calculated above may be significantly reduced if a straight chute is replaced by a staircase profile (Peyras et al., 1991; Chanson, 1994; Degoutte, 2006).

The advantage of this technique is that each step will partly dissipate the energy. In Cemagref's tests on a 1:5 scale model, energy dissipation was quantified for weirs at heights of 3 m, 4 m, and 5 m; 1-m-high steps; and 1:1, 1:2, and 1:3 slopes. Once all the calculations are done, the required basin length starting from the weir toe is:
– for a 1:1 slope: $L_b = 3.6q + 0.9$;
– for a 1:2 slope: $L_b = 3.45q + 1.0$;
– for a 1:3 slope: $L_b = 3.57q + 1.4$ (m and $m^3/s/m$ units) (Peyras et al., 1991).

Apron

A simpler and less expensive option would be to extend the chute with a horizontal apron covering a few metres (the length of the hydraulic jump) and to plan to fix any post-flood damage. The apron, often made of loose riprap, works like a fuse plug. In the event of overflow at a low flow rate, the hydraulic jump will erode the areas around the apron or even dislodge part of it. However, the chute will not suffer any damage and the levee will not be compromised.

Project designers often choose this solution, which is illustrated in Chapter 5 by Figure 5.8.

Spillways with a fuse plug

One obvious method to "minimise the damage" caused by unrelenting efforts to raise levees is to raise spillways with fuse plugs or movable gates. The protection level mentioned above will then become the top of the fuse plug rather than the top of the fixed section of the spillway.

The maximum capacity of the spillway may be achieved for heads between the waterline of the safety flood and the fixed section of the spillway. If H refers to the head above the fuse plug and F is the length of the fuse plug, then the fuse plug will increase the spillway's capacity per metre at a ratio of $[(H + F) / H]^{3/2}$. However, this requires flow to remain free, even once the fuse plug has collapsed. Otherwise, the ratio will be lower.

For example, assuming that flow remains free, with a 50 cm head on a 1 m fuse plug, the ratio will be 5.2. It will still be 2.8 for the same head on a 0.5 m fuse plug.

The length of the spillway can then be divided by this value for the same protection level. Or the protection level can be increased for the same spillway size.

As we can see, a more complex installation will give the designer more leeway. We will examine the different types of mobile or fuse plug devices in Chapter 5.

The required modelling

Overview of the types of hydraulic models

In one-dimensional (1D) models, we consider the flow to be straight and the waterline horizontal in each section. Transversal slopes are ignored. These models do not allow us to estimate velocities in the riverbed.

We differentiate single-reach models from branched models that take tributaries into account and meshed models that take multiple branches into account.

Two-dimensional (2D) models allow for simulation of flow in a plan view and the inclusion of irregular geometry and obstructions in the floodplain. The velocity at each point is a vector equal to the average velocity along a vertical axis. We must know the surface distribution of the floodplain's roughness.

1D storage cell models are intermediate models. They take into account storage cells in the floodplain that are based on topography (hillsides, levees, embankments). These storage cells store water and do not contribute to the flow; as such, they have a uniform water level. These storage cells are connected to the riverbed or floodplain and even to each other through a weir or opening. These models calculate faster than 2D models and can take into account the role of floodplain attenuation while being structured like 1D models. They should therefore not be called 1D+ models, which could mislead non-specialists: they are indeed 1D models.

Three-dimensional (3D) models are useful for better understanding flow and velocities near a complex structure.

A single overall model or two uncoupled models?

When a spillway overflows, the flow will be diffluent. With a return spillway, flow will be meshed.

Two types of flows interest us here:
– Flows in the leveed river, which may or may not overflow the spillway.
– Flows downstream of the spillway through a channel, unconstructed floodplain, or a confined area (flood expansion area or protected area).

As long as flow over the spillway weir is free, the flow beyond the spillway will have no impact on the river's flow. Both calculations can be uncoupled to facilitate modelling. An initial calculation in the river will provide input data for the second calculation in the floodplain downstream of the spillway.

However, if flow over the weir is submerged, uncoupled models are no longer possible and an overall model must be used. In practice, a wide range of peak flows and flood durations must be tested. This is because strong or long floods could submerge the weir. As we previously mentioned, the optimum size of a spillway in a flood expansion area is with a submerged configuration. An overall model is therefore required. However, in a protected area, spillway operation during a danger flood can occur in free flow, since we do not want excessive inflow. Nevertheless, since the project designer will run a simulation of the opening of a breach during a study of risks, an overall model might still be useful.

In both cases, uncoupled and therefore simple calculations may help determine the feasibility of a project and test new design possibilities.

Steady or unsteady models?

It goes without saying that we are interested in unsteady conditions: the flood expansion area or protected area fills gradually according to the shape of the flood hydrograph affecting the spillway. An unsteady model is therefore essential here.

However, it may be worthwhile to use a steady model of the leveed riverbed for flows equal to the peak flow of the protection flood. This will help determine the profile of the spillway crest.

1D or 2D model?

If we need to perform coupled calculations for both flows in the riverbed and the discharge area, a 2D model is clearly necessary (see example in Figure 3.24).

However, for uncoupled calculations of just the leveed riverbed, a 1D model may be used if the leveed riverbed has a relatively consistent section (without sudden convergence or divergence). Weir flow must be free across the entire structure; otherwise, we cannot set the flow decrease rule for the riverbed.

For uncoupled calculations in the floodplain downstream of the spillway, a 1D model may be used if the flow passes through a channel or is discharged into a floodplain with a regular shape in which the project designer can predefine the flow direction.

In any event, we have already seen that we must necessarily take submerged flow into account in flood expansion areas. This may not be necessary for protected areas, but the modeller will have to anticipate potential breaches that are most frequently associated with submerged flows. As a result, a coupled calculation will always need to be performed by including the consequences of diffluence.

If we only need to pre-size the spillway in a flood expansion area, we must almost always consider submerged flow conditions. A 2D model will typically be required. But for a protected area, pre-sizing using only free flow conditions is possible.

If the flood expansion area or protected area has a simple geographical configuration, it merely serves as storage and no true flow will occur downstream of the spillway. The area will simply be filled through storage cell hydraulics. In this particular case, a 1D model with storage cells may suffice. But even with uncoupled calculations, a 2D model is preferable if the area downstream of the spillway is slightly uneven, causing flow in multiple directions. A 2D model should also be used if the project designer is unable to reasonably predict the flow direction. And a 2D model is useful for studying the consequences of overflows and the evacuation of a population threatened by flooding.

In closed protected areas or flood expansion areas, water flows first before being stored. When loading starts there is a flow at certain velocities. Then, heading gradually begins downstream and the area fills with water just like a reservoir and then functions as a storage cell. We can see that a storage cell model may come close to reality in this case. It could therefore be useful for pre-sizing. Nevertheless, a 2D model remains critical for calculating water flow velocities in the area to assess the risk of ground erosion or the risks to local inhabitants.

Lastly, 2D geometrical modelling of the spillway remains very difficult as sensitive numerical issues may come up. A better solution would be to model the spillway using a flow law at each abscissa, including any potential submersion.

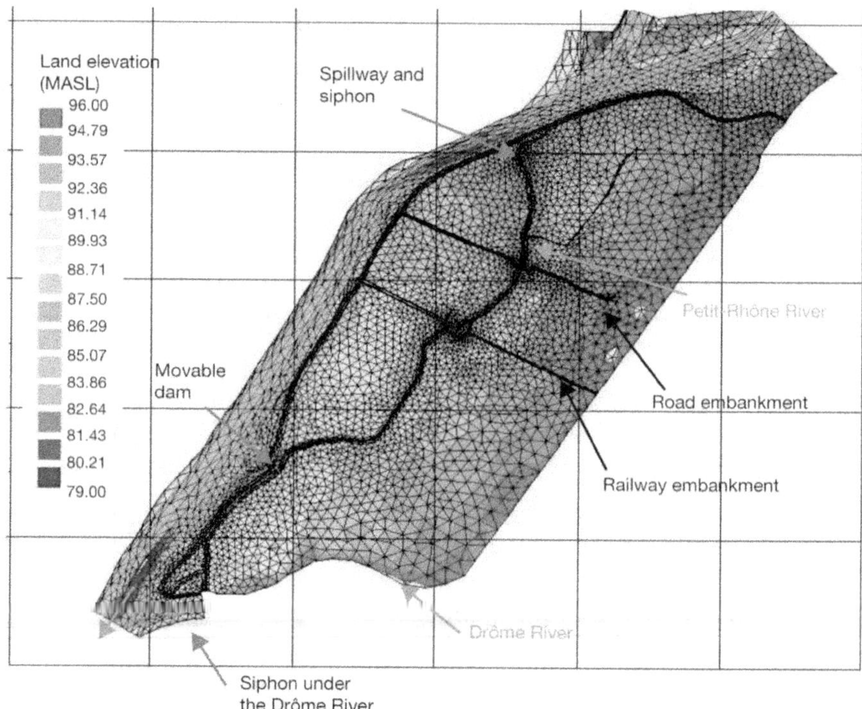

Figure 3.24. Mesh and topography of the CNR structure in Baix-le-Logis-Neuf (TELEMAC-2D code implemented by CNR). The Printegarde plain (flood expansion area) has been split into three storage cells separated by road and railway embankments.

In conclusion, unsteady 2D models should be used as a general rule. The 1D storage cell model should be limited to simple configurations with realistic horizontal water flow contingencies and when we do not need to calculate velocities in the flooded area. Even if a 1D storage cell model is used, it will only allow for pre-sizing and will need to be followed with a 2D model to perform an overall verification and to potentially refine the project parameters.

If the calculation can reasonably be uncoupled, the designer may use a 1D model in the leveed riverbed and another in the floodplain under the spillway, as long as the geometries are simple and regular.

To learn more about this topic, please refer to the methodological guide for hydraulic studies (Cemagref and Cetmef, 2007).

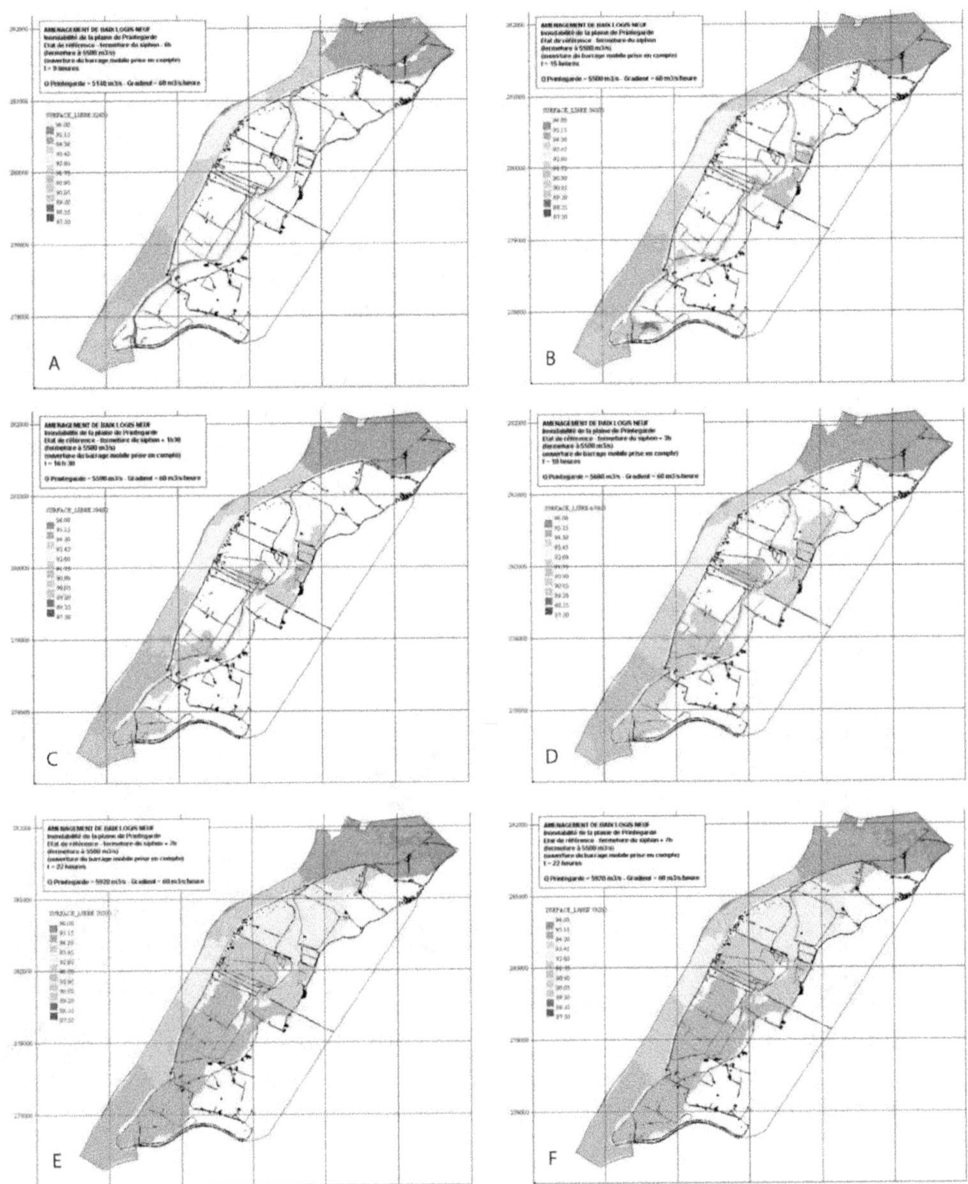

Figure 3.25. Gradual filling of the Printegarde flood expansion area. (Source: CNR)

Example of 2D model outcome

We illustrated how the Printegarde flood expansion area (modelled in Figure 3.25) gradually fills with water for a flood with a rise gradient of 60 m³/s/hour. Photo 3.1 shows that all three storage cells shown in Figure 3.24 were heavily loaded in October 1993.

This example clearly shows that the Rhône River levee was designed to be flooded, hence the "overflow levee" designation used by CNR. For the flood gradient tested, overflows over this levee start at 5,680 m³/s. They would start at a lower flow if the rise gradient of the flood were lower. Overflows will occur in the downstream storage cell where a water cushion has formed thanks to the spillway coming into effect and the siphon closing. Overflows will also occur in the upstream storage cell but without a water cushion. Above 6,220 m³/s, the mobile dam is opened and water levels will balance out between the downstream plain and the Rhône River and the overflow levee will begin to operate in the opposite direction, entirely downstream of the downstream storage cell.

Photo 3.1. Printegarde flood expansion area during the Rhône River flood in October 1993. The red arrows point to the Petit Rhône River (not the one in Camargue). (Source: CNR)

4

Spillways and river geomorphology

Gérard Degoutte
Patrice Mériaux
Yann Queffélean

This chapter addresses the influence of morphological changes in the watercourse on spillway performance and the potential influence of spillways on the morphology of leveed watercourses.

We will also cover the specific case of torrents (slope > 6%).

Overview of sediment transport

As a reminder, there are two means of sediment transport in river channels: bed load, which plays an important morphological role, and suspended load, which plays a more limited morphological role.

Bed load refers to the movement of particles along the bottom via rolling, sliding, or saltation. Bed load transport displaces particles, but they must have already been detached from the bed. This movement is provoked by drag and load-bearing capacity. It is only when the flow becomes highly turbulent that particles are carried away as suspended load, despite their weight. These particles may then travel long distances without ever touching the bottom. A particle displaced via bed load transport may shift to suspended load when the vertical component of the turbulent velocity is higher than settling velocity. Suspended load transport is therefore due to turbulence. It is more intense near the bottom of the riverbed, where the concentration of suspended sediments is higher.

Bed load transport comprises sand, gravel, and blocks, while suspended load transport comprises sand and silt.

Suspended materials come from the riverbed and interact with it. Materials smaller than those on the bottom, which come from the catchment area or arable soil in the riverbed, are carried away as wash load and do not settle on the bottom of the riverbed (except in dam reservoirs). They are evenly carried away as part of the flow and are less than 0.06 mm in diameter. These fines may settle on the floodplain when the water level falls, or remain trapped in bank vegetation. They generally do not play a morphological role (Degoutte, 2006).

Torrents are also relevant here since they are frequently leveed in the alluvial fan that forms in the outlet onto the main valley. In torrents (slope > 6%), bed load transport plays a critical role and has a strong impact on liquid flow. The flow level is significantly higher than flows comprising just water. Furthermore, within the same range of slopes, the fluid is no longer considered Newtonian.[9] Lastly, some torrents may contain debris flows with a high concentration of mud and stones that can travel very long distances and reach far higher than a "liquid" flood.

Overview of morphological changes in watercourses

We will not examine leveed watercourses just yet. If the watercourse has not attained a dynamic equilibrium, its longitudinal profile may change over time, either deepening or rising. This deepening process is called erosion. Raising the profile is known as aggradation. Furthermore, both these processes can occur in an upstream or downstream direction. There are four types of interactions in a longitudinal profile: backward erosion, progressive erosion, backward aggradation, and progressive aggradation. Both forms of erosion (backward and progressive) are much more frequent than aggradation. They are sometimes collectively called "incision". Despite their similarity, these mechanisms are very different.

Backward erosion results from the riverbed being lowered (excavation, weir destruction, a meander being cut off, etc.)

Progressive erosion results from a deficit of materials carried away via bed load transport. This deficit may stem from the extraction of aggregates from the river bed, a weir or dam trapping materials in the riverbed, afforestation that traps materials in the catchment area, the significant inflow of unloaded water, etc.

These overall mechanisms should not be confused with local erosion or deposits. A river section will reach dynamic equilibrium when the erosion/deposit balance is null. In this case, erosion and deposits are generally reversible and may be viewed as part of the riverbed's respiration.

Morphological changes in leveed watercourses

Building a levee on both banks of a watercourse has two strong morphological effects:
– On a watercourse in dynamic equilibrium, this leads to incision of the riverbed, especially if the levees are high and the space between them[10] is reduced.

---

9. The shear strength of a Newtonian fluid is proportional to the velocity gradient that is applied to it. This fluid will flow even under very low stress, whereas a non-Newtonian fluid will only flow above a specific stress threshold.

10. That is, the unprotected floodplain on each bank.

– Suspended load materials will settle in the space between the levees as thick deposits, which increase as this space is reduced (Photo 4.1). Of course, these deposits are thicker in watercourses where the suspended load is high. They vary according to the frequency at which water flows into the unprotected floodplain.

Photo 4.1. Unprotected floodplain aggraded with silt deposits (yellow mark) on the left bank of the Grand-Rhône River

When a levee breach occurs, the flow that is released will carry four types of materials and deposit them in the floodplain over varying distances:
– Materials from the levee itself
– Materials from the erosion pit that forms at the breach toe
– Some suspended load materials
– Materials the river continues to carry, as in watercourses where the bottom is higher than the banks (see Photos 4.2 and 4.3)

Photo 4.3. Alluviated house on the left bank of the Doménon River, 10 days after the breach. (Photo: LTHE)

Photo 4.2 The Doménon River after an emergency repair of the levee, view from downstream. The yellow arrow points to the flooded and alluviated house in the Summer of 2005, also shown in the second photo. (Photo: R. Tourment, INRAE)

These volumes can be quite high. During the 2003 Rhône River flood, the Claire Farine breach released water into Petite Camargue (right bank of the Petit-Rhône River):
– 2.5 to 85 cm of sand over 800 m^2
– 0.2 to 2.5 cm of silt over 3,800 m^2 (IRSN, 2004)

Another example featuring a torrent is the Doménon, a tributary of the Isère River that flows down from the Belledonne mountain range. A levee was built on its alluvial fan (Photo 4.2). In August 2005, a high flood caused the leveed water bed to alluviate and overflow. The left bank levee gave way over 30 m, and large deposits of material piled up to a height of 1 m, with fine deposits reaching a thickness of 40 cm (Source: LTHE). Photo 4.3 shows coarse deposits around a house at the foot of the breach.

Keep in mind that building a levee on a river will cause incision, which will interfere with the spillway's performance if not planned for. However, even a river that has been leveed in recent decades has not necessarily reached equilibrium and might still be in the incision stage when the addition of a spillway is being considered.

Aggradation of an unprotected floodplain with silt could make it difficult for water to feed a spillway that is not much higher than the unprotected floodplain.

When considering whether to build a spillway, designers should also remember that by preventing a breach, a spillway could help prevent both violent floods and significant sediment deposits.

The spillway's influence on sediment transport (riverbed)

A spillway has local morphological impacts by interfering with streamlines and an overall effect by diverting flow. We will first discuss the latter by considering a watercourse in dynamic equilibrium (being neither aggraded nor degraded).

The spillway's influence on the longitudinal profile

Since bed load carries particles from the bottom of the riverbed, the spillway will have no direct impact on it and will not divert coarse sediment flow – only suspended fines. Downstream of the spillway, the following situation occurs:
– Reduced liquid flow.
– Unchanged sediment flow downstream.
– The watercourse therefore has a relative excess of sediment discharge that will produce coarse material deposits that start near the spillway and spread downstream. This is progressive aggradation, which is not limited to that spot.

The extent of this progressive aggradation will depend on how high the divertible flow is and how frequently it is diverted. It will therefore vary according to the spillway. A diversion spillway in a channel will produce a great deal of divertible flow. When a spillway discharges water into a flood expansion area, the divertible flow may or may not be significant depending on the size of the area. However, with a spillway meant to protect a small protected area, divertible flow will be limited, as will diversion.

If progressive aggradation occurs, it is in the form of a tilting slope: a steeper slope is needed to carry such a high sediment load with less liquid flow. Progressive aggradation that occurs downstream of the spillway will raise the level of the bottom along the spillway. As a result, backward aggradation will occur upstream of the spillway if the slope is the same.

In turn, this aggradation will affect the spillway's hydraulic performance by raising the waterline. This increase could have harmful effects if the project designer does not take it into account. Earlier and stronger overflows might occur in a protected area.

A flood expansion area may fill with water too quickly and therefore be less effective when actually needed. The project designer must therefore assess this aggradation, which is based on the amount of diverted flow and the frequency of the diversions.

For instance, a spillway that would only start discharging for a 100-year flood would have just a minor influence on the longitudinal profile, unlike a spillway that discharges for a 10-year flood. If an assessment of the spillway shows that it has a significant influence, the project designer could suggest preventive measures such as topographical monitoring of the river bottom and the possibility of raising the weir to offset this aggradation after a few years.

However, the impact of aggradation downstream of the spillway may have adverse effects not limited to the spillway's operation. The longitudinal profile of the riverbed will tilt, creating a steeper slope.

Downstream of the spillway, there are two positive impacts (reduced flow and increased slope) and one negative impact (raising of the bottom). In general, we cannot predict what will happen. The effects are positive far downstream of the spillway but may be negative near the spillway (Figure 4.1).

Upstream of the spillway there is only a negative impact: increased elevation of the riverbed bottom. The waterline slope and flow remain unchanged. Except in the waterline connection area, the situation worsens (Figure 4.1, circle on the left).

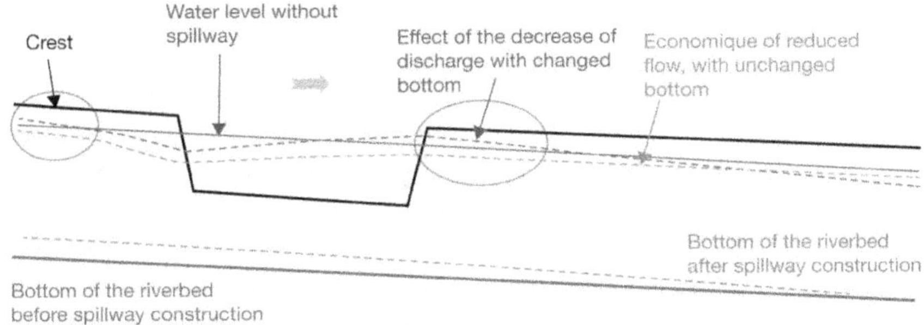

Figure 4.1 Effect of a spillway built on a leveed river with a high bed load and dynamic equilibrium before the spillway was built. Far downstream of the spillway, there is a hydraulic benefit (less liquid flow). Close to the spillway, a morphological impact may occur (reduced sediment transport capacity). The same impact would occur upstream of the spillway. The spillway may therefore have negative effects in the circled areas.

We initially assumed that the river is in dynamic equilibrium before the spillway is built. However, if it is aggrading, the problem would be even worse (see above). On the other hand, if the river is in the incision stage, aggradation and incision would occur in opposite directions. If incision is due to a levee built a few decades ago, construction of the levee and spillway might very well lead to incision since the spillway would operate less often than the levees themselves.

The spillway's local influence

A spillway also has a local influence when it operates since it will change the direction of waterlines and produce eddies. This local phenomenon is similar to diffluence, such as between the Petit-Rhône and Grand-Rhône Rivers, or at the entrance of oxbows. At the downstream end of the diffluence where the Y forms, recirculation currents develop at a very low speed. Hence the deposits (Rosier, 2007). These deposits may extend downstream through a wake effect. They become visible at the end of a flood and look like islands (Figure 4.2 and Photo 4.4) or peninsulas. Vegetation will gradually cover the deposit if flooding does not occur for several months, and will make it even harder to displace again.

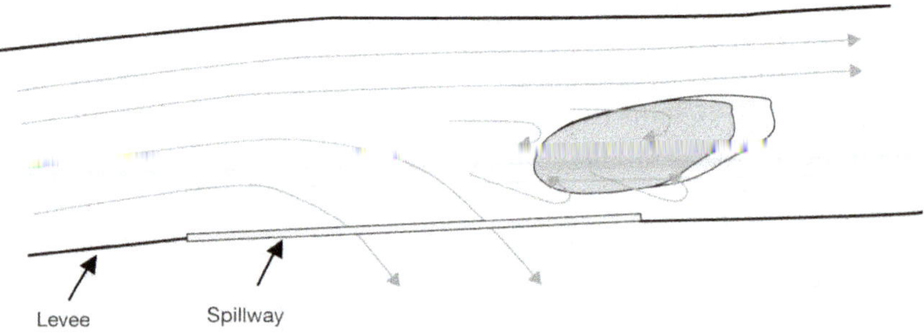

Figure 4.2. Deposit settlement downstream of a diffluence.

In summary, two different mechanisms have the same effect near the downstream end of the spillway: aggradation of the bottom. This will raise the waterline along the spillway and increase both the overflowing nappe and the diverted flow, which usually has harmful effects.

On a balanced watercourse, it is important to monitor any changes in the river-bed and regularly remove any vegetation that could gradually overrun the bed, for two reasons:
– To facilitate the movement of material during floods.
– To reduce the effects on the waterline.

If deposits become too thick, they should be removed (by displacing the material further downstream if the river lacks sediments).

On a balanced watercourse, the presence of a spillway will cause progressive aggradation downstream and backward aggradation upstream. It will also form deposits in the immediate surroundings. These mechanisms coexist and will lead to the spillway being used more often than expected. Downstream, the raising of the river bottom could reduce the benefit of the spillway or even cancel it out. Upstream, it will most certainly have aggravating effects.

However, these effects should be put into perspective for spillways that are seldom used, such as for greater than 50-year floods or spillways that divert low flows.

Luckily, hydraulic requirements and morphological consequences have a similar effect of encouraging diversion only for infrequent floods. And in protected areas, the diverted flow is typically very low compared to the watercourse's morphogenetic floods. This is another reason to limit the morphological impact.

Nevertheless, we do recommend always examining the morphological aspects and sediment transport.

Photo 4.4 Spillway on the left bank of the Savoureuse River (built by the Pays de Montbéliard Urban Community). The spillway's placement on a concave bank helps preserve a channel between the deposits and the spillway. Vegetation must not be allowed to form on the island that is created.

Sedimentation behind spillways

A spillway that diverts liquid flow will also divert suspended sediment flow. This sediment flow will increase faster than the diverted liquid flow (as the concentration is an increasing function of the liquid flow).

For a diversion spillway in a channel, a simple calculation of shear stress will indicate whether deposits tend to occur in the channel. Typically, a diversion channel has a steeper slope than the watercourse, and will simply move suspended fines when it is relatively narrow and deep. A broad and shallow channel will be subject to deposits and will gradually be covered in vegetation, which will further increase deposits while reducing flow capacity. The weeds in the channel must be cut to maintain flow capacity and prevent any increase in sedimentation. Dredging may also be necessary, depending on how substantial the deposits are. The designer should therefore prioritise the construction of flood channels that are deep enough to minimise or cancel out deposits, or at least estimate the average annual deposits to include this figure in maintenance costs.

If the spillway does not release water into a channel but rather into a flood expansion area or a protected area, the fines will settle. They will either be deposited near the spillway outlet if there is no slope, or near the downstream end if the area is sloped and closed. These areas can be identified through 2D modelling to assess the range of speeds. The accrued volume may be calculated by measuring concentration in the riverbed during periods of high water. Conventional empirical concentration formulas could also be used (Bagnold, etc.). This volume is generally low in a protected area since the diverted flow is relatively low and the spillway is used infrequently. In any event, it is clearly much lower than the volume that would result from a breach. Deposits may form in a flood expansion area over time, but they are not necessarily harmful for farmland.

The influence of bed change on spillway performance

We will now look at the watercourse's influence on the spillway, rather than the spillway's influence on the watercourse. Any morphological changes in the longitudinal profile will inevitably alter the spillway's operating conditions. If bed change is consistent in the area, the spillway will overflow more often with aggradation, and less often with incision. It will eventually no longer function as expected.

A more serious problem is an inconsistent change in the area since the general levee elevation may eventually become inconsistent and the spillways will be used either too often or never. One example is the Loire River levee near Orléans, which overflowed well before the Jargeau spillway – built further upstream after the 1866 flood – went into effect. This levee had been sized to channel the 1825 flood without overflows, but a lot of sand was removed over the last century, leading to significant backward and progressive erosion that tilted the longitudinal profile (Maurin et al., 2004).

The case of an aggrading watercourse

For a river section that is aggrading, one solution is to set up a deposition area that is cleaned regularly. The SYMBHI (Joint Development Association for the Isère River Hydraulic Basins) is considering this option for the Isère River upstream of Grenoble, as it is planning to create several flood expansion areas (Figure 4.3).

The case of an eroding watercourse

We will now look at the opposite situation, an incised river. If there is backward erosion, the cause must first be addressed if it is artificial. Otherwise, a solution could be to install a grade control weir to stabilise the longitudinal profile (Figure 4.4).

However, if the incision is due to progressive erosion, a grade control weir would be useless: the problem would be displaced downstream of the weir and the incision would continue. Nevertheless, a flow control weir could be installed downstream

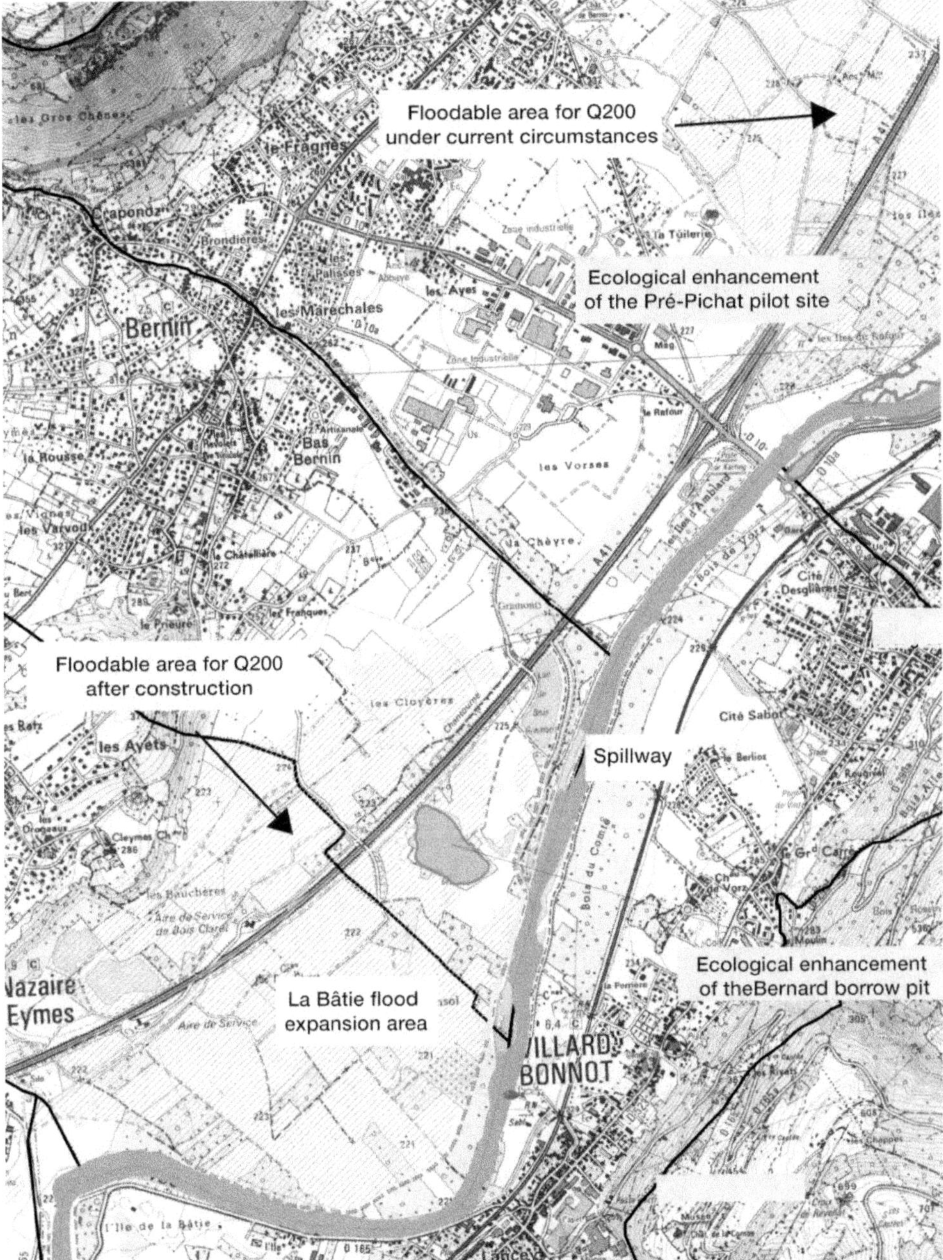

Figure 4.3. Diagram of a flood expansion area project in La Bâtie, on the right bank of the Isère River, with an upstream spillway. The projected deposition area spreads over 2 km upstream of the spillway, up to the Brignoud bridge (northern end of the diagram). (Source: SYMBHI)

of the spillway (Figure 4.5) and, if necessary, several consecutive weirs. Reducing the hydraulic gradient of the waterline will reduce the upstream sediment transport capacity and adjust the new longitudinal profile to a watercourse carrying a smaller bed load. This situation is typically found downstream of large reservoirs or trapped gravel pits. The spillway profile must take the presence of the weir into account. Both solutions are similar but have different designs.

Depending on the circumstances, and above a specific height, these weirs must be designed so that fish can pass. Although these weirs are initially designed as grade control structures with no falling water (Figure 4.4), they do function as true barriers.

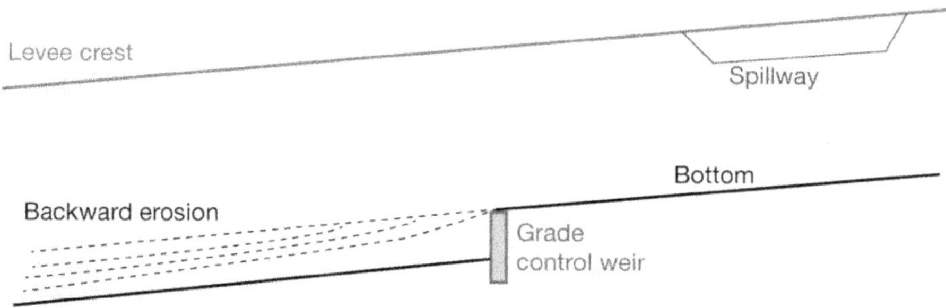

Figure 4.4. Grade control weir to stabilise the longitudinal profile of a watercourse with backward erosion.

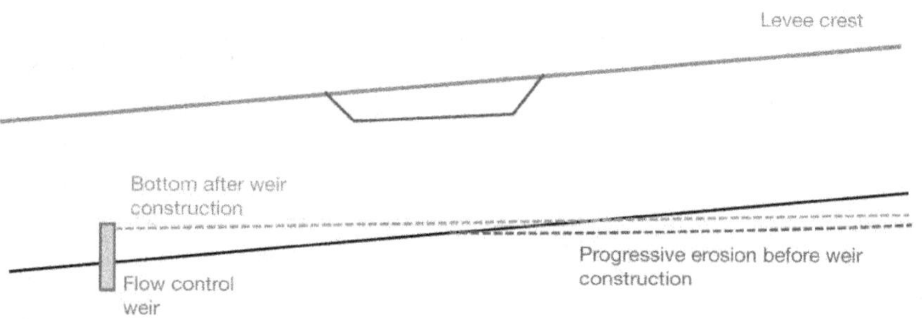

Figure 4.5. Flow control weir to change the slope of a watercourse and adapt it to a lower bed load.

Torrents

With river levees, installing a spillway introduces slow and controlled flooding into the area that had been protected until then. By lowering the waterline both upstream and downstream, the spillway also delays generalised overflow of the levee.

With a levee to protect against torrents, spillway operation is harder to predict since many aggravating factors (sediment transport, debris, large blocks, floating matter, etc.) may significantly hinder the spillway's performance. In particular, unless serious preventive maintenance is performed, severe jams are very likely to occur in sometimes unpredictable places during floods that affect the spillway. Flooding of the Valdaine River after a violent storm in the summer of 2002 and the one of the Belledonne range in August 2005 have confirmed these effects. Not to mention the disastrous jam in the Morge River in June 1897 that blocked the watercourse at the entrance to Voiron, when 7 to 8 m of water accumulated.

To adequately position a spillway, the potential changes in the bed (both reversible and irreversible) should be considered. Only looking at floodwater levels could lead to serious failures since overflows could occur elsewhere before the flood level reaches the crest of a poorly positioned spillway. Because of the supercritical flow regime, the spillway will not lower the upstream waterline.

It should also be noted that, due to the cone shape of the area, a spillway on a torrent levee will not typically create a stilling water cushion.

For torrents and torrential rivers with bed load (with a mainly liquid phase), the utility of installing a spillway on a protection levee should be evaluated on a case-by-case basis, taking the previously mentioned limits into account. As we mentioned earlier in this chapter, releasing some of the flow outside the leveed bed will directly impact the torrent's transport capacity and may contribute to its alluviation (since overflowing liquid flows no longer contribute to bed load transport). As a result, the spillway will function differently than expected, and may paradoxically worsen overflows.

For torrents likely to have mudflows, installing a lateral spillway would likely be ineffective since a debris flow may move into a channel on its own due to its rheological properties. In any case, attempting to separate the upper layer of a debris flow would be fruitless, since flows separated from the main stream would immediately become alluvial ridges (debris flow has a "threshold fluid" and needs to reach a sufficient height – depending on its rheological properties and the slope – to be able to flow).

Conclusion

In conclusion, for **rivers with high bed loads**, caution is required and sediment transport conditions must be thoroughly analysed to determine whether the watercourse has reached dynamic equilibrium or is in an incision or aggradation stage.

For rivers in dynamic equilibrium, the spillway may function more often than expected due to downstream aggradation. Or the spillway may reduce the benefit of the diversion both upstream and downstream if it frequently diverts a substantial flow. However, these effects may be very limited if the spillway is only used for exceptional flows and, therefore, for short periods.

When these effects are insignificant, or at least acceptable, it is better to install the spillway on a concave bank if possible. In any case, installation on a convex bank is out of the question.

For a watercourse that has not reached equilibrium, a spillway could turn out to be useless if it is never used, and even harmful if it is used too soon. Solutions such as weirs or deposition areas are possible, but the latter would incur maintenance costs.

In the end, both morphological conditions and hydraulic efficiency suggest building structures that would only overflow during relatively exceptional floods.

For torrents, which are located on alluvial fans, building a spillway is not necessarily a good idea as they have little practical effect on debris flows. As for liquid floods, their operation might be difficult to predict considering a strong risk of disruption due to heavy bed load or extensive floating matter.

5

Civil engineering design for levee spillways

Paul Royet
Gérard Degoutte
Stéphane Bonelli

This chapter examines safety spillways in protected areas and spillways that pass flows into flood expansion areas or flood canal.

A spillway is typically a fixed structure that functions passively. It usually comprises a weir that controls the amount of water that flows into the protected area or flood expansion area, followed by a chute and an energy dissipator. It may also feature a movable or fusible raising device placed above the fixed spillway weir.

Photo 5.1 shows a levee spillway on the Rhône River with a weir made of concrete beams embedded in the levee, followed by a chute made of stepped gabions and a hydraulic jump basin to dissipate energy.

Photo 5.1. Comps spillway (Gard department) with the downstream face made of gabions.
(Source: DDT 30)

The weir's conveyance is always an important criterion for flood expansion areas and often for protected areas as well. The discharge head is never very high since there is no point in discharging water too frequently or creating levees with very high crests. As a result, most spillways need to be very long, which can be prohibitive because of the site's conditions or financial considerations. As a result, solutions are needed to maximise the flow per linear metre. One frequent option on dams is a movable weir that can be lowered at a strategic moment, releasing a flow rate per metre that is much higher than with a fixed weir. These movable weirs are adjustable devices such as flap gates or tilting fuse gates that first appeared in 1995. For levees, fuse plug embankments were installed much earlier, on the Loire River in the 1870s. Instead of tilting, they eroded and therefore had to be rebuilt. Another solution to increases a spillway's linear discharge without using a complex movable or collapsible device is the labyrinth weir, which has been used on dams since the 1950s and became even more common in the 1970s. In 2005, an alternative to the labyrinth appeared, the piano key weir (or PK weir).

This chapter will first address fixed weirs, which are simpler and more common whether they are in straight or labyrinth form. We will later examine fuse plugs and movable weir devices.

Safety criteria for levee spillways

Since a spillway is critical to the levee's safety and the efficacy of its flood attenuation, the spillway itself must not put the levee at risk.

The first aspect to consider is the contact between the body of the earthen levee and the overflowing structure made of another material: concrete, concrete riprap, gabions, etc. Just like with dam spillways, the idea is to limit the flow of water at the contact between these two structures and to prevent contact erosion that could lead to piping. This is always an important issue, even more so when the material lining the spillway structure is permeable, such as gabions or riprap.

A second aspect is the ability of the spillway's material to resist loads related to the discharge (and sometimes floating debris). In the case of reinforced concrete, the resistance level is generally sufficient for levees given that nappes are always moderate. Designers have therefore chosen to use cheaper materials, such as concrete riprap or gabions. In this case, the thickness of the nappe must be considered, even for levees, to prevent any damage or excessive strain on the material. If the nappe is thicker, which is often the case in flood expansion areas, and in all cases with collapsible devices, gabions should not generally be used. If the collapsible device is adjustable or a fuse, concrete must then be used because of its strength and the flatness that is required.

Since few levees have spillways or the spillways have not yet been used, we lack feedback about problems that could emerge.

The first bypass spillways on the Loire River, such as in La Bouillie, have failed many times. Yet we have very little information about these failures and their sources. The initial structures were certainly simple and possibly paved earthen notches that were not very resistant to discharge.

In 2002, a spillway on the Vidourle River in Lunel breached during a flood. Even without the spillway, the levee would have failed. There is no detailed record of this breach and it may even have occurred on the "regular" section of the levee next to the spillway. This hypothesis should encourage engineers to make sure their elevation settings are correct. When the structure's capacity is exceeded, overflow should not be concentrated on the levee's "regular" section, and above all, right next to the spillway. The breach might also have occurred due to contact erosion between the levee soil and the structure (underneath or on one side). This shows the importance of proper and careful design and construction (anti-piping shields, filters, good ground compaction, post-construction profile verification, etc.).

Principles of fuse plugs or movable devices

With a collapsible device as shown in Figure 5.1, the area will be protected from flooding up to the crest of the device z_2, whereas the fixed section is at level z_1. In the event of a high flood, in comparison with the same structure without a movable or fusible device, discharge on the spillway will be delayed and will occur closer to the flood peak. As soon as the movable or fusible device is removed, the spillway will be able to divert large quantities of water from the river, and therefore lower the waterline both upstream and downstream. This is why a movable or fuse plug structure is generally very effective in hydraulic terms. Compared to a fixed spillway built at level z_2, conveyance will be increased at a $[(z_3 - z_1) / (z_3 - z_2)]^{3/2}$ ratio, provided that flow remains free. Compared to a fixed spillway built to level z_1, flooding of a protected area is delayed and the capacity of a flood expansion area is increased. These benefits should be weighed against potential dysfunctions.

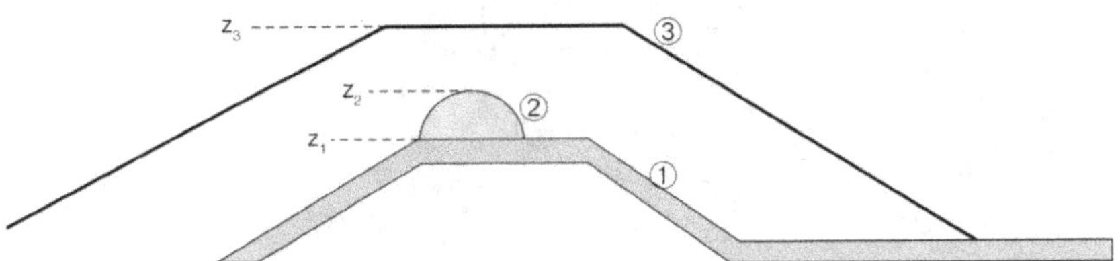

Figure 5.1. Cross-section of a spillway fuse plug.
① Cross-section of the fixed spillway ② Fuse plug or movable device ③ Cross-section of the straight levee section.

The plan shape of the levee spillway

Concerning the spillway's plan alignment, the simplest solution is aligning the spillway with the levee. This spreads the nappe across the plain, ensures homogenous right-of-way, and offers visual continuity. But unless the spillway is covered by a footbridge, it will obstruct access to the levee crest during floods. Since spillways are typically long, building a footbridge that will not prevent floating debris from flowing over would be costly. No examples of this exist in France. However, the Bonnet-Carré spillway on the Mississippi River (Photo 1.12) does have a road above it since it is required to operate the needles.

Maintenance vehicles may drive across the spillway outside flood periods if both connections between the weir and the levee crest are gently sloped.

Another solution might be a more compact structure such as a "duckbill" spillway with three overflowing sides (Figure 5.2). It protrudes from the levee on the riverside. If a footbridge is added, since it is much easier to install in this case, foot traffic can continue along the levee crest, including during flooding. The drawbacks are a concentration of flow towards the protected area (which will require specific protection against erosion), marked visual impact, and more expensive foundations and sidewalls than for aligned spillways. This solution is only feasible for short spillways. We do not know of any spillways that use this system.

However, an underpass closing structure under the Arles-Tarascon railway track illustrates the duckbill principle well (Photo 5.2). To prevent high floods in the Rhône River from flowing through the underpass, a duckbill-shaped earthen embankment was built. The intent was not to build a spillway but to delay or prevent floods from flowing under the railway tracks. During the December 2003 flood – which was just over 100-year flood level at peak flow – the embankments, a bit lower than the railway tracks, began to overflow and erode.

Photo 5.2. Aerial view of the duckbill-shaped earthen embankment built along an underpass on the Arles-Tarascon railway. (Photo: SYMADREM)

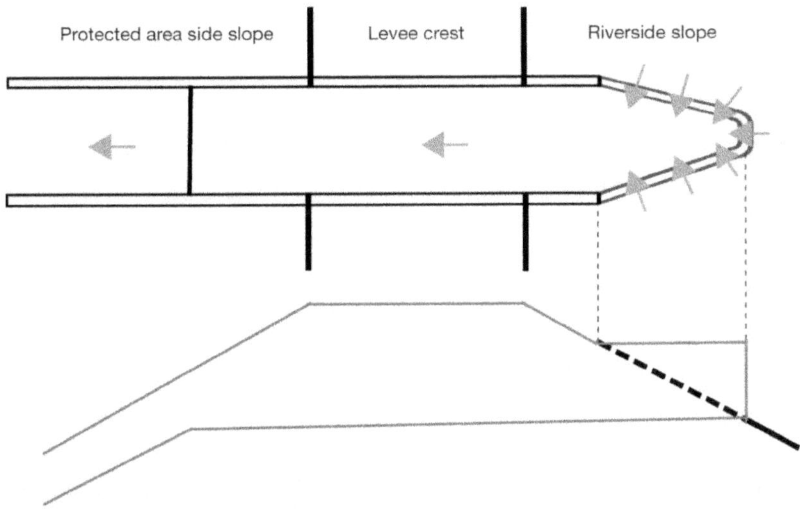

Figure 5.2. Plan view and longitudinal cross-section of a duckbill spillway.

Another alternative for long spillway weirs is a labyrinth spillway weir. This is a sequence of small duckbills that create a zigzag pattern (Figure 5.3 and Photo 5.3). The weir length is extended as a pecked line that is typically four times longer than the weir length in a straight line. If the head reaches half of the wall height, the discharge rate per metre is twice as high. This technique has been used many times in the USA to renovate old embankment dams (Hincliff et al., 1984). There are no examples of this for levees, but it could be useful if the desired conveyance requires a very long structure that may not be possible in certain locations.

Photo 5.3. Labyrinth spillway on the Fontaine des Gazelles dam (Algeria).

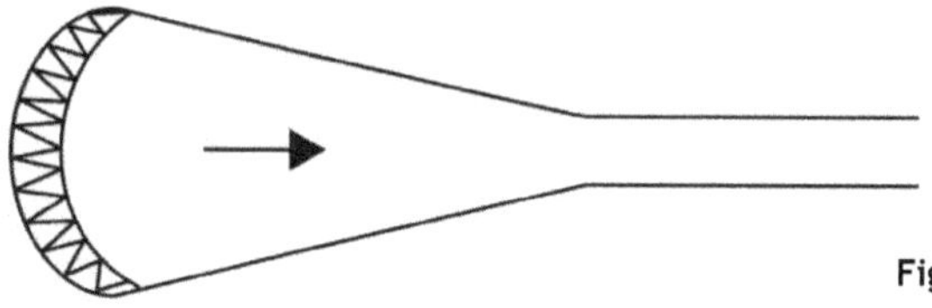

Figure 5.3. Plan view of a labyrinth weir.

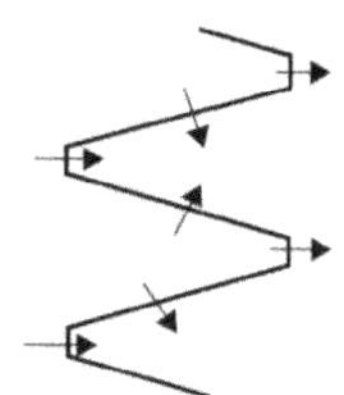

The piano key weirs (or PK weirs) recommended by Hydrocoop further increase the weir's conveyance for a given area (Lemperrière et al., 2003).

PK weirs have a labyrinth-like crest shape, but unlike labyrinths, their walls are not vertical (Photos 5.4 and 5.5). The walls of the upstream section are tilted downward so that flow will plunge into a cavity known as an outlet ①; the walls of the downstream section are tilted upstream and the cavity is then called an inlet ②. As a result, the structure will have a substantially reduced upstream-downstream footprint thanks to two fairly symmetric upstream and downstream overhangs. This kind of system is also very effective in hydraulic terms. The outlet produces downward suction that will remove any hindrance in the upstream part of the labyrinths thanks to the convergence of the nappes. The inlet produces upward suction and will promote ventilation of the nappe thanks to the downstream overhang.

Conveyance may be more than three times higher than for a well-designed straight weir with the same footprint. The foundation of this structure should be carefully designed to prevent differential settlement.

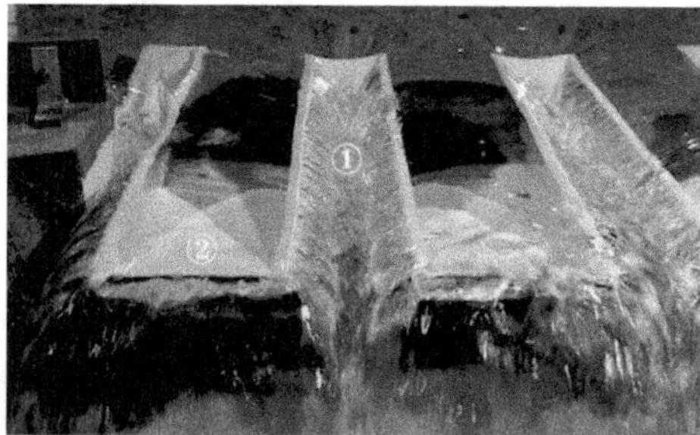

Photo 5.4. Scale model of the PK weir on the EDF dam in Goulours. (Source: EDF-CIH)

Photo 5.5. Construction of the PK Weir on the EDF dam in Saint-Marc. (Source: EDF-CIH)

About 15 PK weirs have been built in France. The first was in Goulours in 2006, followed by others in Saint-Marc, l'Étroit, les Gloriettes, and Record (Laugier et al., 2018). In total, there are now more than 35 PK weirs around the world (the USA, South Africa, Australia, Sri Lanka, Scotland, etc.). Alongside France, Vietnam in particular has been a driving force in the promotion of this type of spillway.

A global PK weir register was created in 2017 (Erpicum et al., 2017) to list all completed or ongoing PK weir projects. The world register or pk weirs, including project sheets, is freely accessible on the University of Liege's PK weir portal at: https://www.uee.uliege.be/cms/c_5026433/fr/world-register-of-piano-key-weirs-prototypes

Several PK weirs have been installed on embankment or rockfill dams (the Peach Tree dam in the USA, Black Esk Reservoir in Scotland, Loombah dam in Australia etc.). As with labyrinth weirs, this technique might still be useful if space is limited. They have proven to be effective even in the presence of floating debris (Laugier, 2007; Pfister, 2013). Tests in Vietnam showed that conveyance was still good even if the spillway was submerged from downstream (Ho Ta Khanh et al., 2011; Dabling et al., 2012).

We will now put aside labyrinth-shaped structures to focus on straight weirs.

Types of straight weirs and constituent materials

Various designs are possible when installing a weir on a levee. The main criterion is the height of the nappe that might flow over the weir:
— If it is low (from one to several decimetres), a simple levee lining may be used (Reno mattress, riprap armouring, etc.) to provide resistance to erosion.
— If the nappe on the weir is likely to be high (metric order of magnitude), solutions such as gravity structures (rigid works made of standard concrete, roller-compacted concrete, or solid masonry) are preferable.
— For low levees that protect stakes of little importance, simpler solutions such as controlled erosion embankments can be used.

Gravity-type spillways

These structures typically have a triangular or trapezoid weir that is extended on the landside by a stilling basin. When building a new structure, two techniques can be used:
— Unreinforced conventional cased and vibrated concrete: the upstream face (riverside) will be vertical and the downstream face will have a batter of around 1H:1V to minimise the volume of concrete.
— Roller-compacted concrete (RCC): casing should be avoided; requiring a weir with a symmetric profile and an upstream and downstream batter of around 1H:1V.

The choice between these two solutions will mostly depend on the quality of the foundation and economic considerations. Regarding the foundation, various studies and recent construction projects have shown that it is possible to build a rigid structure on a loose foundation, as long as it has a symmetric profile (which would result in a rather homogenous distribution of stresses on the foundation, whether empty or filled with water). On the contrary, a profile with a vertical upstream face will require a rigid foundation. In financial terms, RCC is preferable to standard concrete when at least tens of thousands of cubic metres are needed.

Massive spillways such as gravity structures are heavy works that are likely to provoke substantial settlement in absolute terms, as well as differential settlement. A study of the foundation's compressibility should be performed, and any sections where breaches have previously occurred or with paleochannels should be closely monitored.

Massive spillways made of standard concrete

In this solution, given the price of standard concrete, the volume should be kept to a minimum. The typical weir has an upstream vertical profile with a downstream batter of around 1H:1V (Figure 5.4). The weir crest has a rounded profile for better hydraulic efficiency (discharge coefficient of about 0.45 to 0.5).

If the concrete has been properly poured, the structure should last for a long time as concrete resists repeated and substantial nappes without erosion.

The most important aspect is the quality of the foundation. Given the highly asymmetrical distribution of stresses on the foundation, it must be mechanically sound (using rock or consolidated alluvium) to prevent any differential settlement that could lead to cracking. Considering the high hydraulic gradient (about 1) that will occur under the structure during floods, the foundation should also be treated to prevent any risk of erosion (spuds, contact grouting, etc.).

In the longitudinal direction, construction joints should be built every 15 m or so to allow for thermal shrinkage in the concrete. These joints do not necessarily need to be sealed.

The sidewalls that provide a connection between the weir and the regular section of the levee should be either made of reinforced concrete or use sheet pile walls driven into the substratum if possible. But if space is required for a maintenance path that can be used outside of flood periods, these vertical sidewalls should be replaced by more gently sloped concrete ramps or concrete riprap.

Depending on the nappe likely to flow over the weir, the downstream face should be extended by a stilling basin or a simple concrete apron ending with a concrete spud or a sheet pile cut-off.

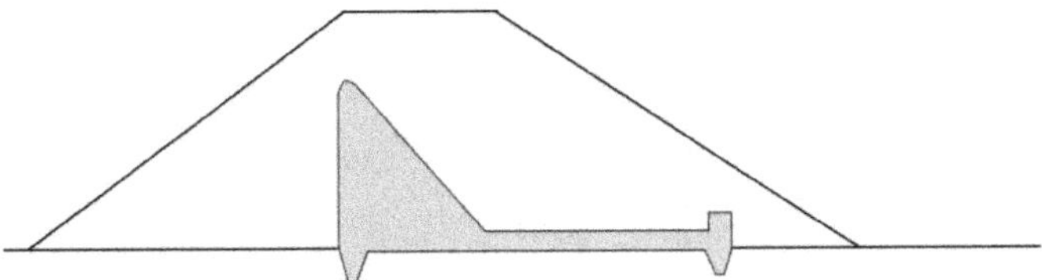

Figure 5.4. Profile of a standard concrete weir.

Vehicles cannot drive across this structure since its crest is too thin. However, ramps linking to the crest of the overflow parts of the levee can be built on either the watercourse or landside. The landside is preferable to ensure that vehicles can pass even when the unprotected floodplain is flooded, as long as water is not flowing over the spillway.

Massive spillways made of RCC

This spillway is a massive weir made of RCC that is extended by a stilling basin also made of RCC. RCC is a proven material in dam construction that is suitable for overflow structures:

– The first RCC structure built in France was a weir (in Saint-Martin de Londres, Hérault department).

– The RCC technique is often used around the world to build temporary submersible cofferdams.

– Several dams, mostly in the United States, whose spillways were deemed insufficient were equipped with additional RCC spillways placed on the embankment.

With this design, stresses remain low everywhere on the structure, making it possible to use rough RCC (hard embankment) in low quantities made of rough alluvium (Londe et al., 1992).

A typical RCC profile (Figure 5.5) is symmetric with a batter of 1H:1V both upstream and downstream, with no formwork required. To simplify construction, the weir crest should be around 3 m wide. This results in a massive structure that is appropriate for high levees (above 5 m) and long spillways. As watertightness is not crucial, no joint treatment sealing is required. Given the operating return period of such a spillway and the satisfactory behaviour of RCC during overflows, no standard concrete lining is needed. Any erosion that might occur during overflow is usually acceptable.

The reinforcement of the foundation is designed to prevent piping from occurring at the point of contact between the RCC weir and the alluvial foundation. The idea is to drive a sheet pile cut-off through the alluvium deep enough to limit the hydraulic gradient and prevent the risk of internal erosion. The cut-off depth may be determined using Lane's principle or other criteria in the literature comparing the maximum hydraulic gradient with an acceptable gradient value according to the foundation materials.

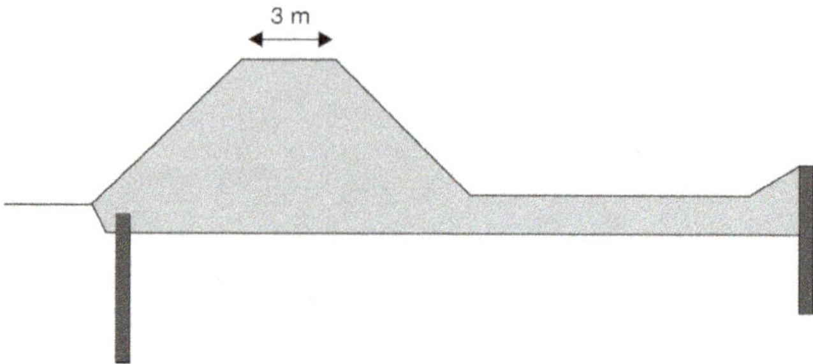

Figure 5.5. Typical profile of an RCC weir.

The sidewalls that connect the weir and the straight section of the levee can be built from sheet pile cut-offs embedded in the foundation.

Energy will be dissipated downstream of the weir by a catchment basin also made of RCC ending in a sheet pile cut-off a few metres below the apron.

Unlike a standard massive concrete spillway, the levee crest will remain accessible (except during overflows) if the sidewalls are gently sloped. Otherwise, ramps may also be used.

Covered notch spillway

In this case, the weir comprises a lining placed on the crest and the downstream face after partial excavation of the levee. The device will also include a catchment basin at the landside toe. The structure will be laterally connected to the non-overflow levee via similarly covered ramps or limited by sheet pile cut-offs that serve as sidewalls in the levee and guide walls downstream.

The lining may be flat and made of reinforced concrete slabs, riprap or concrete riprap, or Reno mattresses. Or it could be made of stepped gabions.

For aesthetic purposes, a flat lining may be covered with a grassy topsoil layer that will quickly erode in the event of overflow and will therefore need to be restored after each overflowing flood. Unfortunately, this type of lining hides the spillway (especially given that it can remain in place for more than a century without being used) and we may forget about the related risk downstream. Photo 1.4 shows how easy it is not to notice a spillway.

These structures are relatively light and far less likely to cause settlement than gravity spillways. However, it is important to watch out for the slightest settlement or a faulty altimeter setting on the levee. We want to avoid centralised overflow on the levee once the spillway is saturated, especially next to the spillway.

Reinforced concrete slabs

A lining made of cast-in-situ reinforced concrete slabs, without upper formwork, is certainly the most complex solution that requires understanding the compressibility conditions of the levee and its foundation. Limited maintenance is needed

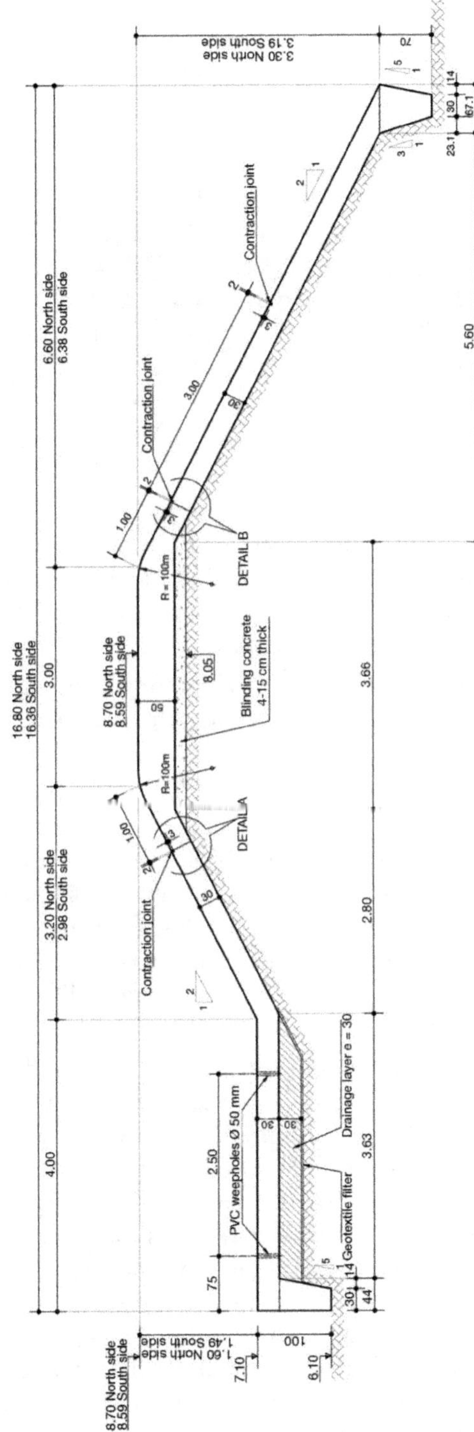

Photo 5.6. Reinforced concrete slabs being placed on the downstream face and apron of the Lattes diverter spillway. (Source: BRL-i)

compared to concrete riprap or Reno mattresses, but energy dissipation is not as good. Photo 5.6 and Figure 5.6 show the Lattes diverter spillway previously presented in Chapter 1. In this example, the top width (3 m) allows vehicles to pass.

Figure 5.6. Cross-section of the 150-m-long spillway in Lattes. The Lez River is on the right. (Source: BRL-i)

Concrete riprap

A surfacing material made of concrete riprap is a more common solution, especially for very long structures. With this technique, no construction joints are required. It has a maximum recommended hydraulic load of 1 m and can carry overflows with a velocity of up to 8 m/s.

This technique was used to build the 900-m-long Aramon spillway in 2003. The levee crest is 1.10 m above the crest of the weir. The downstream face, crest, and upper part of the upstream face are made of 200–400 mm concrete riprap. A geotextile lining provides the contact between the riprap and the embankment.

The Aramon spillway overflowed immediately after it was completed, during flooding of the Rhône River in December 2003. The spillway functioned for about 5 hours with a maximum nappe of 15–20 cm (Mallet et al., 2004). Relatively abundant seepage occurred at the levee's downstream toe due to faulty sealing between the geotextile lining and the concrete riprap. This problem was subsequently fixed by injecting a grout curtain through a line of holes drilled on the crest at 0.5 m intervals.

On the right bank of the Gardon River, close to the confluence with the Rhône (on its right bank), a fuse plug spillway eroded during the December 2003 flood. It was replaced in 2006 by a 30 m fixed spillway with a weir 1.7 m lower than the levee crest. The downstream face is made of 400–800 mm concrete riprap. In Figure 5.7 and Photos 5.7 to 5.10, we can see the reinforced concrete beam at the top of the spillway. The beam makes it possible to calculate a more accurate height-discharge relation than a weir made of concrete riprap. In addition, it provides a seal by blocking flows that would pass through riprap without concrete, or where the riprap meets the ground. The riprap is placed on a gravel drainage bed. Drain outlets help release any (static) uplifts generated by leaks and reduce dynamic pressure generated by internal flows. An ejector will introduce a depression generated by the overflow velocities (part of $V^2/2g$) (SOGREAH, 1962).

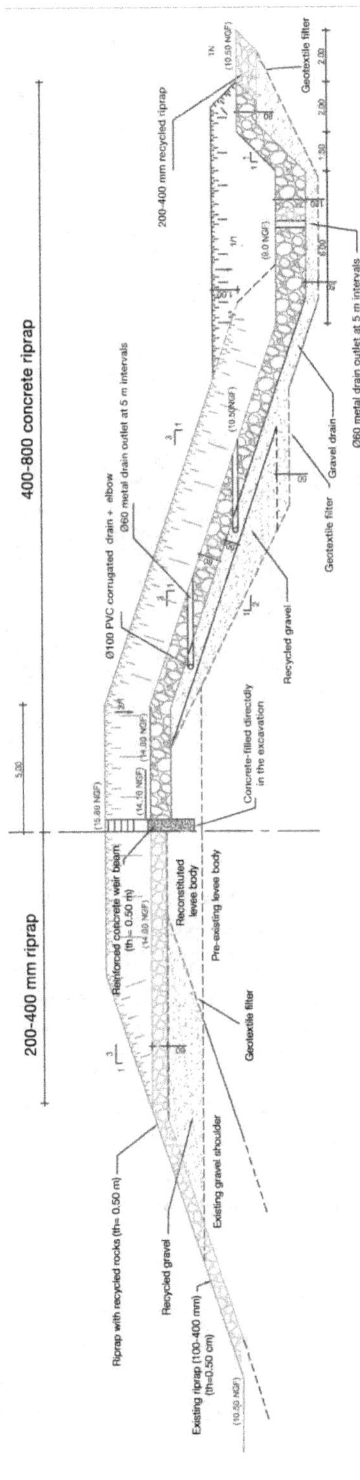

Figure 5.7. Typical cross-section of the Comps spillway. The Gardon River is on the left. (Source: CNR)

Photo 5.7. Comps spillway, installation of outlet drains. (Source: CNR)

Photo 5.8. The crest beam formwork has been stripped off. (Source: CNR)

Photo 5.9. Placement of downstream riprap. (Source: CNR)

Photo 5.10. Downstream face and completed stilling basin. (Source: CNR)

Feedback on the behaviour of concrete riprap during floods has been rather positive. INRAE performed a thorough inspection of small flood-attenuating dams in Nîmes after two high floods on 9 September 2002 and 8 September 2005. The spillways on these structures are made of 400–800 mm concrete riprap, generally with an apron with a 1:2 slope. These structures, which were loaded by nappes of at most 0.56 m, functioned well and showed no sign of cracking.

CACG also gave good feedback on dams dating back to the 1990s that were designed for irrigation. These structures were lower than 12 m, with loose foundations, spillway chutes with gentler slopes (1:2), and smaller riprap (100–300 mm). A few small stones have loosened but are easily repaired.

Riprap

Weirs made of riprap are a well known and common technique in rivers. These structures are entirely made of blocks, where little sealing is required.

The technique we are presenting is slightly different since we are referring to a riprap lining. However, the rules for sizing blocks to resist currents are the same.

A block placed on the bottom of the riverbed will remain stable until the velocity of the current exceeds entrainment velocity. We use Isbash's formula here to calculate d, the median diameter of the blocks based on U, the velocity of the surrounding current:

$$d = 0.7 \frac{\gamma_w}{\gamma_s - \gamma_w} \frac{U^2}{2g}$$

For a very turbulent flow, this value will be twice as high.

For block density of $\gamma_s / \gamma_w = 2.6$, the formula is $d = 0.022U^2$ or $d = 0.044U^2$ based on turbulence conditions.

In this formula, d is the intermediate value for all three sizes of an ellipsoid-shaped block, which is the smallest square mesh of a screen that would let the block through. Specifications for riprap are often provided as a mass or nominal diameter. The nominal diameter is the cube root of the block's volume, namely the edge length of a cube-shaped block with the same volume. If the block were a perfect sphere, the nominal diameter would be

$$d_n = \sqrt[3]{\pi / 6} \cdot d = 0.81d.$$

On average, it is about 0.85 times the actual diameter. In this case, the block volume (d_n^3) will be about $(0.85d)^3 = 0.6d^3$, and the block mass in tonnes will be

$$0.6 \, \gamma_s / \gamma_w \, d^3.$$

A geotextile lining must be placed underneath the riprap as a filter to prevent water from dragging fines away from the levee body. It also helps protect against erosion on the levee surface.

Since riprap must be installed without tearing the geotextile lining, a layer of small riprap should be placed between the lining and large blocks. To block particles from the soil skeleton, the geotextile lining must come in close contact with the ground at all points, meaning that a flexible geotextile lining should be chosen and intermediate-sized granular material should be placed in between to ensure proper distribution of the stress caused by the riprap. This granular material should obey filtering rules for riprap to ensure it is not dragged away by the current. Another benefit of this transition material is that it prevents sunlight from reaching gaps in between the blocks, which is critical given that geotextiles are sensitive to UV radiation.

Reno mattresses

Reno mattresses (or gabion cushions) are well suited to structures with low hydraulic heads not exceeding 0.7 m at the crest and with a maximum velocity of 6 m/s. As with concrete riprap, we strongly recommend installing a concrete beam (Figure 5.8, Photos 5.11 and 5.12).

The Reno mattresses stones should be placed so they cannot be washed away by the current. If there is no netting, stones will detach according to a Shields parameter higher than a critical value of 0.05. With caging, this critical value is doubled (Simons et al., 1983).

We need to check that

$$\frac{y \cdot i}{1.6d} \leq 0.1$$

(i = the slope of the discharging face, y = flow depth on the sloped part, d = the median diameter of the mattress stones). Once all these calculations are done, we need to ensure that flow depth does not exceed 50% of the median diameter of the mattress stones for a slope of 1:3. This ratio increases from 50% to 70% for a 1:4 slope and 90% for a 1:5 slope. This verification is absolutely necessary and may require adjusting the spillway length, the slope of the face, or the size of the stones.

A geotextile lining underneath the entire gabion section is essential and the gabions should be placed carefully to avoid tearing the lining. The geotextile will serve as a filter to prevent water from washing fines away from the levee body and to protect against erosion on the levee surface.

Photos 5.11 and **5.12.** Lunel spillway on the Vidourle River. On the riverside (on the right in the left-hand picture and on the left in the right-hand picture), the Reno mattress is covered in canvas before sodding. On the right, detail of a notched section. Note the concrete beam embedded in the levee to serve as a weir. (Source: Gérard Degoutte)

Stepped gabions

Stepped gabions may also be used as an alternative to Reno mattresses. They will improve energy dissipation (Peyras et al., 1991). The structure shown in Figure 5.9 and previously in Photo 5.1 was built in 2005 by the town of Comps. It is 40 m long. A concrete beam embedded in the levee stands 20 cm above the gabions to serve as a weir. The vertical sidewalls are also made of gabions. A geotextile lining is required, as with Reno mattresses.

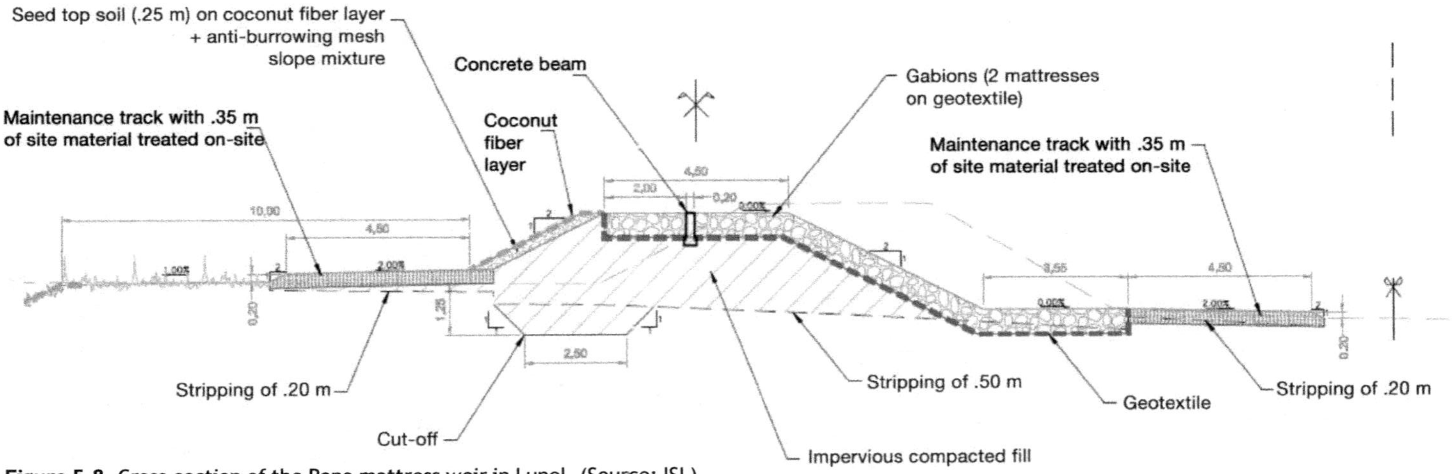

Figure 5.8. Cross-section of the Reno mattress weir in Lunel. (Source: ISL)

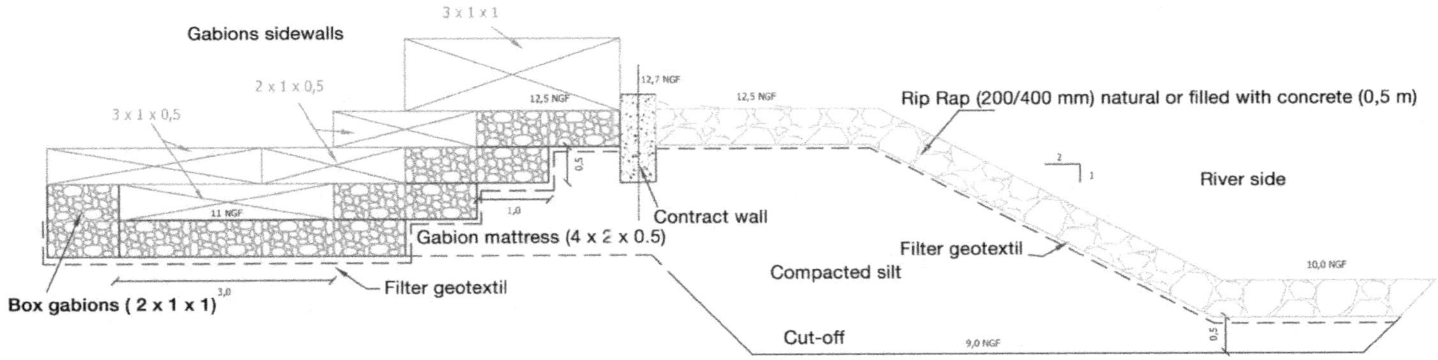

Figure 5.9. Cross-section of the stepped gabion weir in Comps. (Source: DDT 30)

The potential of lime-treated soils

Soils have been treated with lime for centuries to facilitate their use and improve their native properties to build infrastructure.

Introducing a small amount of quicklime (2 to 3% by dry weight) into silty or clayey soil changes its physical properties (reduced plasticity since lime causes the flocculation of clay particles, drying thanks to the heat released during the hydration of CaO, reduced sensitivity to swelling and shrinkage, and increase of the bearing capacity after compaction). Lime makes plastic and/or wet soil much easier to handle and use. In addition, the soil's long-term mechanical properties may also be improved by incorporating higher amounts of lime (the dosage depends on the soil's clay content). This "stabilisation" is possible thanks to a so-called pozzolanic reaction between silicates and aluminates in the clay and hydrated lime in the presence of water to form a binder that gradually hardens.

In France, lime-treated soils are widely used in transport infrastructure (embankments, subgrades and pavements). They are less used in hydraulic works, probably because of a lack of knowledge about their hydraulic properties (hydraulic conductivity, erosion resistance, durability in a hydraulic context). However, in the 19th century, lime-treated clay was sometimes used in hydraulic works (canals, dams). One example is the Cusset levee, built a century ago, which is still performing well. More recently (late 1980s, early 1990s), lime-treated chalk was used to build the Fond-Pignon dam (37 m high) designed to store the sludge excavated during the construction of the Channel Tunnel. More than 1.8 Mm^3 of chalk was treated with 2.5% quicklime to reduce its moisture content (Barthes et al., 1994). Since the early 2000s, small retention dams (5 m high) have been built with wet silt treated with lime: several in Normandy, and three in La Duchère near Lyon (3% lime). The dams and canals in the Seine-Nord Europe high-capacity canal project will be built with chalk and silt that will partly be treated with lime and/or cement.

In non-European countries, particularly the United States, hydraulic structures treated with lime have been built for several decades. The Friant-Kern irrigation canal in California, built in the 1950s with heavy clayey soils, experienced damage to its banks caused by the swelling of these materials (cracking, sliding, erosion). The US Bureau of Reclamation repaired and strengthened the canal in the 1970s. Clay banks were treated with lime (4%) and rebuilt to improve their stability and ensure resistance against erosion from water flowing through the canal. More than 40 years later, this technique remains effective (Howard et al., 1976; Herrier et al., 2012). There are also other examples of applications to correct or protect poor, erodible, or dispersive soils:
– The United States: the Mississippi River levees (6 m to 12 m high) (Gutschick, 1978); the Los Esteros Dam (67 m high) (Mc Daniel, 1979); the McGee Creek Dam (49 m high) (Knodel, 1987).
– Swaziland: the Mnjoli Dam (42 m high) (Forbes et al., 1980).

– Australia (Phillips, 1977).
– Thailand (Cole et al., 1977).

Since the beginning of 2010, actions have been undertaken in Europe, and particularly in France, to evaluate the resistance to internal and surface erosion of soils treated with lime. INRAE evaluated concentrated leak erosion using the Hole Erosion Test on Rhône River levee material. These tests showed that resistance to concentrated leak erosion with clay loam increases from 2 m/s to 10 m/s (2% lime, 14-day curing time). Jet erosion tests conducted by the US Army Corps of Engineers on levees in New Orleans also showed significant improvement in resistance to surface erosion following lime treatment (Gutschick, 1978; Benahmed et al., 2012; Herrier et al., 2012). Another important experimental earthfill levee was built along the River Vidourle in July 2015 as part of the DigueELITE R&D programme, in cooperation with the contracting authority EPTB Vidourle. One section of this levee (3.5 m high) was made of low plasticity silt (PI = 5) treated with 2% quicklime while the other section was not treated. The two sections were tested for surface erosion by applying steady artificial overflow. Two series of tests were performed, 1 and 2 years after construction, using the following procedure: tests lasting between 4.5 hours and 17 hours, progressive increase of the water flows up to 600 l/s/m, overflow height of 30 cm, maximum water velocity at the toe of 5 m/s. The following conclusions were made (Nerincx et al., 2018):
– On the crest, erosion of lime-treated soil was 6 to 7 times lower than untreated soil.
– On the slope toe, erosion of lime-treated soil was 5 to 10 times lower than untreated soil.
– A significant erosion pit was created at the toe of the untreated section.
– In the upper part of the slope, erosion magnitude was similar for both zones.
– In the lower part of the slope, erosion of lime-treated soil was 3 times lower than untreated soil.

Lessons learned and the positive results from the full-scale experimental site showcase all the benefits of the technique:
– Available soils can be used even when they are of poor quality.
– The mixing technique helps ensure soil homogeneity.
– Mechanical properties and resistance to erosion are improved.
– The permeability level remains similar to untreated soil when compacted with a padfoot roller on the wet side of the Standard Proctor curve.
– It solves a variety of problems that are likely to occur in levees (animal burrows, roots, internal erosion, surface erosion).

Lime-treated soil can be seen as a new material for hydraulic structures that will open up the possibility of new design techniques. If future tests confirm this potential, one solution could be to treat overflow levees with lime, in combination with standard spillway treatments (riprap, etc.), or even as an alternative. The lime treatment process is also well suited to very long structures or the low submersible dam concept (Lempérière et al., 2017).

Fuse plugs or movable devices

This type of device is nothing new. In the 19th century, Comoy began installing fuse plugs made of sandy soil small embankments on the Loire River levee spillways that are still in place. In the past few decades, gravel fuse plugs have been installed on spillways along the Rhône River. However, observations during floods, laboratory tests on models, and studies of these devices have not yet provided answers to many uncertainties about the fusibility (initiation conditions and erosion kinetics) of these devices (Royet et al., 2004).

To ensure these structures erode or collapse as expected and, most importantly, to better manage kinetics, various fuse plug and movable devices have been designed in the past few decades, including:
– inflatable weirs;
– hydroplus fusegates;
– removable slabs.

The first two types of devices are already commonly used on dams, and inflatable weirs are also used in rivers.

Inflatable weirs

Inflatable weirs[11] are made of 10-mm-thick flexible reinforced rubber tubes. The first inflatable weir was built in the United States, but few people know that this process was invented in 1949 by a Frenchman, Dr Mesnager. In France, there are about five of these structures dating back to the 1960s and two others built in the 1990s. There are several hundred inflatable weirs in the USA and more than one thousand in Japan.

These weirs are inflated by water or air. Water-inflated weirs, which can support high loads, do not seem justified for levee spillways given their limited head (1 m to 1.5 m). In practice, inflatable weirs can be built up to 2 m high and 100 m long. The tube is fixed to the structure with metal flat plates bolted to one or two mooring lines in the weir. The air supply lines, if any, are incorporated into the support beam (Figure 5.10).

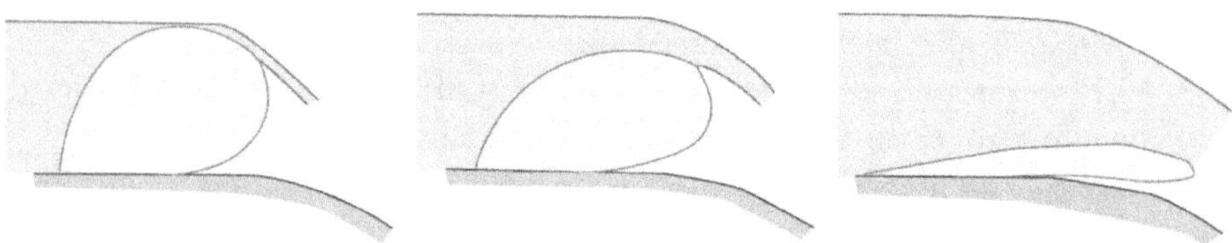

Figure 5.10. Block diagram of an inflatable weir.

---

11. Also (mistakenly) called inflatable rubber "dams". See https://en.wikipedia.org/wiki/Inflatable_rubber_dam

For rivers with slow flood kinetics, we suggest deflating these weirs under normal conditions to prevent vandalism and protect against UVs and various forms of damage by placing the weirs in a covered gutter made from a blockout in the concrete slab. The tube's 10 mm thickness protects it from damage caused by rodents.

When a flood is expected, and if the water level is likely to reach above the fixed weir level, the gutter covers will be removed and the weirs inflated using motor-compressors (this process should be included in flood instructions provided by the levee manager and in the local emergency action plan) (Tagwi, 2015).

Depending on the flood level, managers will choose when to deflate the weirs, thereby flooding the protected area, to avoid overflowing an earthen levee. Deflation can be managed by an automatic control system.

Other than inspecting the membrane for damage and putting the gutter cover back on, no specific action is required to reactivate the device after the flood (which is useful in the event of floods in quick succession).

This solution offers substantial benefits in terms of flexibility and reliability. Experience on river dams and weirs suggest that exposed tubes should last for more than 40 years. In our case, since the tubes are protected and rarely loaded, their lifespan might reach up to 100 years. Operating tests should still be performed every two to five years. These tests can also serve as flood management training exercises.

Inflatable metal-flap weirs

A variation of the inflatable weir involves placing a metal flap on the upstream part of the rubber tube to support the water pressure. This process was created in the USA and patented by Henry Obermeyer. The applications described below pertain to dams but could also be used for levees. One example was built in 2005 on the Meuse River in Villers-devant-Mouzon by Voies Navigables de France (VNF) (Poligot-Pitsch et al., 2007). It comprises three 5.m-long and 2-m-high openings. An automatic control system triggers inflation as the water level increases. VNF built another structure like this in 2010 in Auxonne, to replace a needle dam on the Saône River (Photo 5.13 and Figure 5.11).

Photo 5.13. Auxonne movable dam with 1.3-m-high gates. (Source: BRL-i)

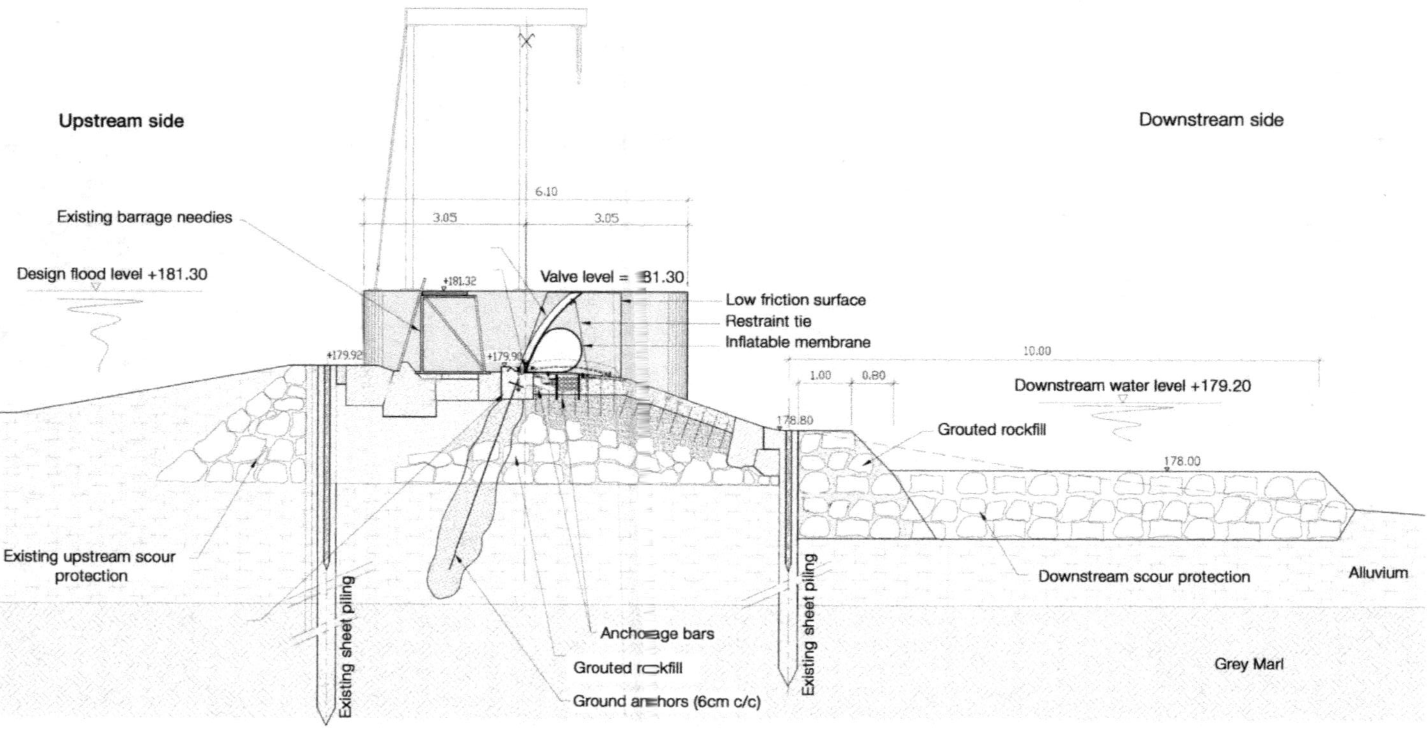

Figure 5.11. Cross-section of the Auxonne movable dam. (Source: BRL-i)

Hydroplus® Fusegates

These are free-standing straight structures placed side-by-side on a weir. They will automatically topple when a specific water level is reached, as predetermined by the flooding of a base chamber (Figure 5.12). For dam flood gates, the number of toppling devices can be adjusted to the flood importance. For levees, toppling gates can be used for a limited range of water levels (5 cm to 10 cm) to ensure the entire flood attenuation capacity of a spillway in a flood expansion area or the entire pre-storage capacity of a protected area is quickly available.

Once toppled, these gates cannot be reused and must be replaced by new ones. Flooding in the protected area will stop when the flood level returns below the spillway's fixed weir.

Photos 5.14 and 5.15 below show an example of this kind of system in a basin designed to attenuate flooding of the Allan River in Montbéliard. Floods were previously discharged through an embankment fuse plug. There was some uncertainty about the operation of this fuse plug due to resistance caused by grass cover and unpredictable erosion kinetics. The Montbéliard authorities decided to improve the structure's reliability by installing Hydroplus® fusegates in 2007. The project was managed by ISL. The new fuse plug spillway has ten gates that are 1.1 m high and 5.6 m long. The first gate will topple for a 100-year flood and all the gates will topple for a 1,000-year flood. The first and last gates will topple with overflows of 27 cm and 37 cm, respectively – that is grading of only 10 cm. As a result, with this gate system, the rarity of the flood event has little effect on the water level.

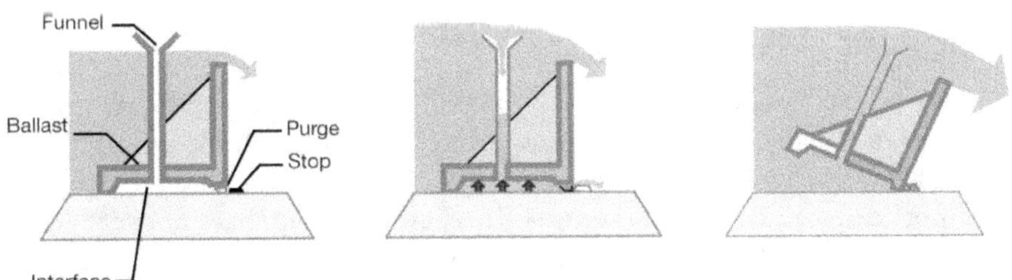

Figure 5.12. The principle of a gradual increase in water level under normal conditions with low overflow; when uplift starts and while tilting. (Source: Hydroplus)

For the Allan River levee, this gate principle was adapted to incorporate the inlet system into a square reinforced concrete tower (Photos 5.14 and 5.15). This tower stills the flow and improves the reliability of the toppling level. It will prevent any malfunctions caused by small floating debris or vandalism. This example is not for a flood protection levee, but rather a flood storage reservoir. However, there is little difference in operation: in one case the water surface of the reservoir is horizontal and flow is frontal. In the other case, which most interests us here, the weir is loaded from the side and the flow line has a gentle slope. Nevertheless, we believe this process can be transposed. These gates were installed on the Huaihe River levee in Anhui Province, China in 1995 (four one-piece concrete gates measuring 5 m wide and 2 m high). But we do not have any feedback on this. The use of fusegates was also considered for many spillways on the Loire River, but none have been built to date.

Photos 5.14 and **5.15**. Fusegates on the Allan River. Top: the gates have been installed; Bottom: the gates are nearing completion with discharging tower and joints between the gates; the Reno mattress is being installed. (Photos: Montbeliard Authorities – CAPM)

Fuse plug embankments

Fuse plug embankments were installed on several Comoy spillways on the Loire River in the 1870s. These are grass-covered sand bars placed on top of masonry spillways as shown in Photos 1.4 and 1.5. Their upstream faces are covered with embedded stone masonry (Figure 1.9). These fuse plugs have not yet been loaded. The Reyran spillway, which is much newer, uses the same principle and has not been used either since the levees were built in 1962 (Photo 5.16). However, the collapsibility of these devices cannot be guaranteed.

However, we can look at real-world experience from the fuse plug spillways on the Rhône River.

The Comps spillway was built on a levee on the right bank of the Rhône River as part of the development of the Beaucaire hydropower plant. In the overflowing part, the levee crest and both slopes are covered by 200–600 mm riprap at 13 MASL. On top of this riprap is a fuse plug device made of two parallel gravel embankments at around 14 MASL. This spillway can be loaded with floods coming either from the Rhône River (from the East) or the Gardon River (from the West).

Photo 5.16. Reyran fuse plug spillway (Var department). (Photo G. Degoutte)

During the Gardon flood on 9 and 10 September 2002, the water level reached 40 cm above the top of the gravel fuse plugs. They eroded slightly without collapsing, probably because the difference between the head on each side of the spillway was too low at the flood peak.

In the days following this flood, the fuse plugs were rebuilt as they were before and scarified. No repairs were made to the spillway itself and the riprap armouring was largely undamaged.

In December 2003, the Rhône River experienced strong flooding. INRAE was able to observe how water flowed over the fuse plug embankment. When the level reached 14.05 MASL, the overflow began to quickly erode a lower section of the fuse plug until it reached the riprap but never extended beyond a few metres. The riprap armouring then began to erode. It took about 20 hours of overflowing with a peak nappe of 0.50 m to almost completely erode the fuse plug embankments. The fact that the fuse plug embankment took so long to breach is probably due to the presence of a double embankment (riprap and gravel).

Fuse panels

A spillway with precast fuse panels is a very simple and appealing concept. It can also be an alternative to fuse plugs.

Vertical panels made of concrete or reinforced concrete are supported by the structure on the waterside to prevent them from tilting towards the riverside. On the downstream side, they are supported by a gravel block designed to resist water pressure as long as there is no overflow. It erodes when the spillway is overtopped, allowing the panels to tilt. We will give two examples of these: a dam project in the United States and a levee installation in Switzerland.

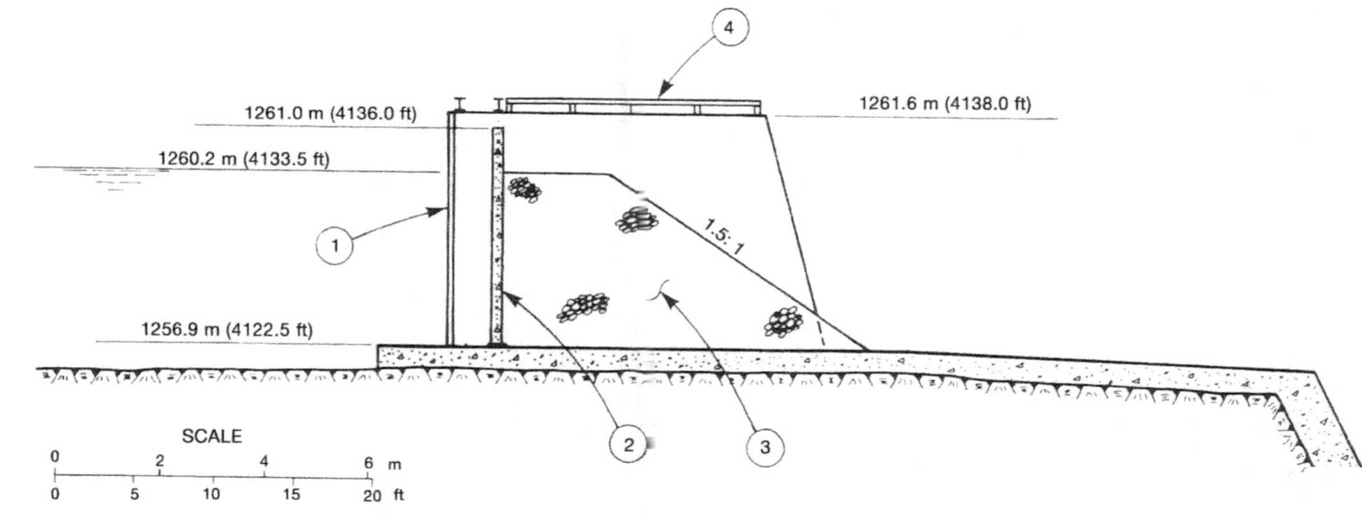

Figure 5.13. Cross-section of the Milner fuse plug spillway.

① Current gate to be removed ② Non-reusable reinforced concrete panel
③ Erodible gravel embankment ④ Bridge and piers (original structure)

Fuse panels on a spillway with a series of openings

The first reference we found in the literature is the renovation of the Milner dam (Idaho, United States) (Pujol-Rius et al., 1991). The existing spillway had a series of fixed-wheel gates. In each sluice, the gate was replaced by a precast reinforced concrete fuse plug panel. The panels are 4.1 m high and lean on the reservoir side of the existing structure. On the downstream side, they lean against an embankment of erodible fine gravel. When water reaches the top of the reinforced concrete panels and overflows, it erodes the downstream embankment. Since the panels are longer have enough support against hydrostatic pressure, the walls tilt and are quickly dragged away by the flow (Figure 5.13). The elevation of the tops of the panels ranges from 1,261 m to 1,261.3 m to prevent simultaneous tilting. Once flooding is over, each opening will be closed by wooden flashboards until the panels and gravel block can be repositioned. Full-scale tests were performed in April 1990 and showed that tilting occurred in less than two minutes.

In the next section, we will describe a similar installation on a levee. However, we wanted to mention this dam installation since it leaves room for a road used for maintenance and monitoring purposes. This could be useful for levees with short spillways. Above all, we have noted the advantage of splitting the fuse plug spillway into groups of slabs (as in this example) rather than slab by slab.

Fuse panels on a continuous spillway

In Switzerland, the Office Fédéral pour l'Environnement (OFEN – Federal Office for the Environment), equipped the Aa Engelberg River with three precast concrete spillways topped with a fuse plug section. The slab toppling principle is similar to the one mentioned above. On the riverside, the slabs lean against a straight concrete weir with a rounded crest. On the landside, they are supported by a gravel bund (Figure 5.14). The slabs are very small and will topple almost simultaneously. This system operated properly during the floods of 21-23 August 2005 (Photos 5.17 to 5.19).

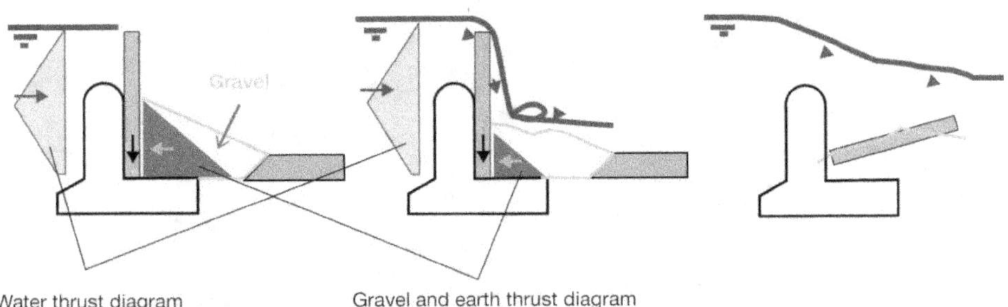

Figure 5.14. Operating principle for the Aa Endelberg River fuse panels. (Diagrams courtesy of OFEN)

Photo 5.17. Fuse panels on the right bank of the Aa Endelberg River. (Photo: OFEN in Rosier, 2005)

Photo 5.18. Fuse plug slabs after the 2005 flood. (Source: Hunzinger, 2007)

Photo 5.19. Aa Engelberg River at its junction with Constance Lake in 2005. We can see two of the three spillways in operation. A secondary levee is under construction to protect inhabited areas (red line). (Photo: the Swiss Air Force in Rosier, 2005)

Even with limited feedback to date, this type of fuse plug seems promising. When the slabs overflow, they send falling water onto the gravel bund, which is swept away by jet erosion. The erosive energy is therefore far higher than with a simple overflow on a fuse plug and control over collapsibility seems to improve. Furthermore, the slab heights can be adjusted so that breaching occurs over a longer period. Of course, once a single slab is gone, we would expect the others to follow by domino effect, even if they do not overflow. But separating walls could be installed to prevent the breach from spreading to a section with slightly higher slabs. We could therefore envision a solution that combines the two previous techniques. The gravel bund would just need to be scarified when overflowing floods occur infrequently enough to allow vegetation to grow. Nevertheless, these solutions remain fairly simple and we cannot accurately determine the breach kinetics, unlike with inflatable weirs or Hydroplus fusegates.

Fuse plug embankment section

This structure is not a simple fuse plug placed on top of a solid weir, but rather a whole levee section. The general idea is to pre-identify a future breach all along the height of the levee and to prevent it from spreading laterally. This simple and low-cost solution should only be used for short levees that protect stakes of little importance. It is not truly a spillway, but we are presenting it here given its low cost and the possibility of choosing the location of the breach. To our knowledge, no constructions of this type exist to date.

The general technique is to adjust the erodibility of the top part of the embankment to initiate a breach when the water level increases. Erosion of the fuse plug embankment is generally caused by overflow (external erosion), but the injection of discharge into the embankment body (internal erosion) also plays a role. Erosion then spreads all along the levee height and probably down into the foundation as well. However, sidewalls will prevent the eroded area from extending lengthwise along the levee. Once flooding is over, the levee section should be rapidly rebuilt since the levee offers no protection in the meantime.

The following design principles are important to remember:
– No important stakes should be located across from the fuse plug section.
– The overflow section should be placed in a location that already has weak spots, such as steeply angled slopes, a narrow crest, or an erodible sandy embankment.
– The levee crest should be lowered by about 50 cm along the preferred overflow length, with an overall V-shaped crest to concentrate the discharge of the initial overflow and encourage breaching. For a very long overflow section, a profile with multiple V shapes should be created to provide several low points to initiate erosion.
– Lateral erosion will be controlled by a sheet pile caisson on each side of the fuse plug embankment that serves as an "abutment" for the structure. These caissons

must remain stable even if the foundation erodes at the breach. To limit the risk of scouring, the caisson should be extended on the landside beyond the levee toe.

This solution is very cost-effective but is less reliable than other options we have presented. The main risk is uncertainty about the levee's collapsibility, which will depend on the type of levee material used and its compactness. As a result, the breaching kinetics and the subsequent flood hydrograph shape in the plain remain unclear. The protected area will also remain submerged much longer as the flood recedes since we must wait for the water level to return to the natural ground level before it stops flowing into the area.

These drawbacks imply that this solution should not be used when the leveed area has human and economic stakes that require a good understanding of flooding kinetics to create monitoring plans.

Conclusion on movable weirs or fuse plug devices

To date, movable weirs or fuse plug solutions have proven their reliability on dams and river weirs, but they are less common on river levees. Using this kind of device on a levee is costly, both in terms of construction and maintenance, given that:

• extensive lengths must be equipped;

• an event that provokes complete collapse or lowering usually occurs more frequently on a levee than a dam.

Recent floods in Europe and China showed that, in times of crisis, it is difficult to operate devices meant to trigger preventive flooding. Local inhabitants rarely accept and will even oppose these operations, sometimes physically. This is a good reason to build devices that do not require human operation.

In parallel, there are many research and development opportunities to design and test other simple devices to achieve a minimum level of sealing as well as high fuse plug erodibility (non-cohesive material that is regularly scarified). The combination of small concrete slabs and fuse plugs seems promising.

Lastly, whatever the type of weir device selected, it is important to ensure the fixed spillway or apron is designed to resist overflowing floods without damage, even if a faulty movable or fuse plug device increases local flow stress.

Continuous maintenance tracks

The overflow section of a levee must comprise a maintenance track on its crest. Another track on the landside toe is also highly recommended. A third on the riverside would also be useful if space allows.

Building a spillway will not affect a track located on the riverside. Nor will it have much impact on the landside if the path is adapted near the stilling basin if any. The main impact will be traffic disruption on the levee crest. Since spillways are typically very long, building a bridge structure will likely be impossible. We know of no such cases in France. The spillway sidewalls should therefore be gently sloped (maximum 20 to 30%), as shown in Photos 1.4 and 1.5 (Loire). Or ramps should provide access to a track on the riverside or, even better, on the landside.

Given that traffic will be disrupted if the flood is higher than the protection level, levee access paths that can be reached from the roads should also be created on both sides of the spillways. This will provide experienced teams with access to the levee crest to perform emergency interventions if it is safe to do so.

Overflow erosion on earthen levees

We described the mechanism of overflow erosion in Chapter 1. In theory, the spillway in a protected area may not delay overtopping of the levee crest for very long. However, it can provide a downstream water cushion to absorb water spilling over the levee and into the plain. Will this prevent breaching or reduce its probability?

We think it would be useful to review the parameters influencing overflow erosion from a civil engineering perspective. We will start with a levee without a spillway.

The parameters that delay erosion are the same that limit loading and improve resistance to loading:
– A crest with a consistent longitudinal profile to prevent concentrated flows;
– A gentle landside slope to reduce velocities.
– A landside slope with a consistent profile to prevent localised eddies, and especially an absence of trees.
– A levee made of cohesive material since stepped erosion reduces water energy;
– A well-compacted levee.
– Very consistent and well-maintained grass cover (inconsistent grass cover can be an aggravating factor by concentrating flows).

These precautions reflect the state of the art for newly built levees, but they may not all be applied to old levees.

What are the benefits of a spillway? The spillway introduces water into the protected area. If it is properly located and designed, a water cushion will form on the landside toe before overtopping begins. This will prevent a jet effect that would create an erosion pit at the levee toe. Ideally, the water levels would be similar on both the riverside and landside. This would ensure that flow over the levee when the crest is overtopped would remain subcritical.

This balance can only be achieved by using properly controlled movable or fuse plug systems. A spillway will not eliminate all risks, but it will substantially reduce their likelihood. And if breaches do occur, they will be less violent thanks to the pre-storage provided by the spillway. Also, people at risk are theoretically protected by the spillway before flooding begins.

More improvements are possible, such as those implemented during the development of the Rhône River, including spillways, overflow-resistant levees that go into effect for rare floods, and non-overflow-resistant levees set for extreme floods (1,000-year flood plus 50 cm in this case). An overflow-resistant levee is designed to be protected by the water cushion and, if necessary, by complementary protection techniques such as riprap or lime-treated soil.

Conclusion

Straight fixed structures are the most common solution. If the required length becomes prohibitive, alternative fixed structures such as labyrinth or PK weirs are possible. Another option is a fuse plug that will release strong linear flows at the right time. The continuity of the access track should be taken into account from the design stage.

For fixed spillways

There are a range of typical solutions. Concrete riprap is the most common. Alternatives such as Reno mattress or gabions can be useful for thin nappes – which is usually the case – and if the risk of floating trees is low. Large riprap is also an option. And lime-treated soils seem promising.

Labyrinth or PK weirs, though they require a more sophisticated design, can reduce the length of the spillway by a factor of three or four.

Movable weir devices

Close attention must be paid to the efficacy of these systems: they must operate at the right time and release the expected flow. There is also a risk that people may try to disable them. This issue should be addressed on a case-by-case basis.

In terms of hydraulic efficiency, inflatable weirs and Hydroplus fusegates allow managers to control the flow that is released. They can be used in both protected area and flood expansion areas.

With erodible fuse plug systems, on the other hand, it is harder to control the kinetics of the volumes released. These systems could be used to feed a flood canal. Fuse plug systems can be used to raise spillways in a flood expansion area if and only if the volume of water that can be stored in the flood expansion area is not a limiting factor. In other words, if and only if the flood expansion area does not risk filling too early and thus not offering any additional flood attenuation if the flood continues. Fuse plug systems could potentially be used to raise the spillway crest in protected areas to provide a better water cushion, as long as a collapse that occurs faster than expected does not endanger local inhabitants.

Fuse plug systems combined with vertical concrete slabs are an interesting option for low fuse plugs, as long as we do not need to know the exact breach kinetics. These systems are fairly simple, but they do require regular monitoring and maintenance to guarantee their erodibility. They are therefore limited to situations in which skilled technicians are available.

6

Emergency management of the levee system

Jean Maurin
Thibaut Mallet
Gérard Degoutte

The purpose of emergency management is to prepare for an emergency, when possible, and get people to safety under the best possible conditions. This chapter will only address levee systems equipped with a spillway and will not cover emergency management on all leveed rivers. We will specifically discuss safety spillways in protected areas since spillways that convey water into flood expansion areas do not require emergency management, as human safety is not an issue. Instead, these areas must monitor the serviceability of essential structures so that flood attenuation occurs as expected.

The benefits and limitations of spillways in emergency management

A spillway offers many benefits in emergency management. The main benefit is critical: a spillway may prevent a levee breach, or substantially limit the flow and volume of water that reaches the protected area. A spillway also limits water levels, velocity, and the duration of flooding. The spillway can prevent a breach if, thanks to the diverted flow, the levee does not overtop, or not long enough to erode the entire width of the crest. The spillway can also prevent a breach even if the levee does overflow by creating a water cushion at the landside levee toe. Nevertheless, the emergency management team must not count on these effects, as the levee's resistance cannot be guaranteed.

In a levee system without a spillway, we do not know the precise protection level: it is therefore the safety level (see Chapter 3). The system will be particularly unstable in an emergency since flooding only occurs when the levee fails, and we cannot know when or where this will happen. Emergency management is quite challenging in this case.

The second benefit of a spillway is just as important: it allows us to know exactly where water will be injected into the protected area, as opposed to the unpredictable nature of a breach. Urbanisation should be prevented in the vulnerable area behind the spillway. The immediate danger area behind the levees should be limited.

The third benefit is that it is fairly easy to determine when the spillway will be activated based on the flood forecast. Emergency management is easier when measures to make people safe are based on a full understanding of the time and location of dangerous events. This benefit is more limited if the spillway features a fuse plug device.

The final benefit is that even if the levee breaches, the impact will be reduced by the presence of water entering slowly behind the levee. And people will already have moved to safety.

Caution! For older levees, the spillway may unfortunately not be the main source of inflow, cancelling the benefits listed above. In other words, the safety level may be lower than the expected protection level. There are two examples of this on the Loire River, one for geomorphological reasons and the other for structural ones:
– The Jargeau spillway on the large levee in Orléans will only go into effect after significant overtopping both upstream and downstream since incision of the riverbed occurred differently near the spillway than on other parts of the levee;
– The La Chapelle-aux-Naux and Le Vieux-Cher spillways that protect the Bréhémont leveed area are meant to operate well before the levee overtops; however, a recent study showed that the levee's safety level is much lower than the operating levels of the first two spillways because of the levee's structural weakness.

In both examples, the spillway no longer serves as a safety spillway and the situation must be handled as with a levee without a spillway where the protection level coincides with the safety level (see Chapter 3). The concept of apparent protection level (Maurin et al., 2012) applies here since **the actual protection level can only be the same as or lower than the safety level.** Even when spillways are present, levee managers should make every effort to determine the actual protection level and perform any necessary interventions to increase it, if necessary.

In addition, the presence of a spillway should not fool levee managers or local populations into thinking there is no longer any risk; it has simply been limited. The spillway might even have a detrimental effect by introducing small amounts of water into the protected area that first seem manageable and do not require getting people to safety. But this flooding could eventually cut off emergency routes if evacuation becomes necessary after the spillway starts operating. This could seriously complicate evacuation efforts. Levee managers should therefore pay close attention to these conditions, especially if the protected area has a slope leading to the hillside (typical of a perched riverbed). The leveed areas on the Loire River are examples of this since the natural ground level at the hillside toe can be several metres lower than it is along the river.

The types of emergencies to manage

In an area protected by levees with a spillway, two types of floods may occur: one controlled by the spillway, and another that is less predictable and flows in through a breach that occurs before or after the spillway is activated. This results in three kinds of risks that we will describe below, ranging from the least to most serious.
– The risk the spillway will overflow. The protected area will be more or less entirely flooded, depending on the flood level. If the protection level is set for a 100-year flood, each year there is a 1 in 100 chance this event will need to be managed.
– The risk that a breach will occur after the spillway overflows; in other words, for a flood higher than the protection and safety levels. For instance, if the safety level is a 500-year flood, each year there is a 1 in 500 chance this event will need to be managed.
– The risk that a breach will occur before the spillway overflows; in other words, the actual safety level is lower than the apparent protection level. Although the flood is lower than in the first case, the inflow of water will be much higher, the water levels and rise gradient will be higher, and flow velocities will be higher. If this event seems possible, and if the apparent protection level is a 100-year flood, every year there will be a more than 1 in 100 chance that this event will need to be managed (for instance a greater than 1 in 30 chance if the safety level is a 30-year flood).

The most serious risk can be ruled out for levees that are properly designed, built (or refurbished), and maintained. The risk of a breach is less likely when there is a spillway, but it cannot be ruled out.

This means that even when levees are in good condition, managers should be prepared to deal with two types of emergencies: a controlled flood, and a sudden flood that occurs after a controlled flood.

Considerations before drafting an Emergency Management Plan

Specific considerations in protected areas

The way the protected area is flooded is critical. If the area is flat and horizontal, which is not common, it will be flooded homogeneously and no distinction needs to be made between people needing protection: they are all equally affected. However, safety measures may vary according to the number of floors in dwellings. We have all seen images on the news showing people being rescued by helicopter from the roof of their house or a supermarket.

If, on the contrary, the protected area has some high ground, this area will be flooded later, or never. These areas may not require protective actions and may even serve as a safe refuge for people living in lower areas. However, a distinction needs to be made depending on whether this high ground can be accessed via routes that remain dry.

In a protected area that is not flat, preferential flows may enter the area and first endanger local inhabitants or people driving through.

We already mentioned the case of a valley in which the riverbed is higher than the floodplain earlier in the chapter. The lowest area can be relatively far from the spillway and the levee. The initial overflow volumes are concentrated in this area and could end up completely isolating the rest of the zone if it is not evacuated beforehand. This situation will be even more problematic if residents are unaware of the trap and slow to seek safety.

An even more dangerous situation is when an area protected by a ring levee is entirely surrounded by water, with the leveed riverbed on one side and the inundated floodplain on the other side (see Figures 1.2 and 1.3). In this situation, the quality of the forecast is critical.

When considering the area's total surface, the number of inhabitants, and the type of road network, another piece of essential information is required: the time needed to get all inhabitants to safety once the warning has been issued. This time may vary depending on whether the flood occurs during the day or at night and whether school is in session.

> An emergency management plan will depend on the topography of the protected area, the height of the buildings, and whether the roads are flooded or not. A 2D hydraulic model is invaluable for assessing the extent of the areas at risk, the order in which they will be flooded, the direction of flow, and the ability to move around the area, particularly to go from the flooded area towards dry hillsides. The emergency management plan must also include the time required to get people to safety in the area or sub-areas.

Considerations for the flood to be managed

The velocity of the flood is an essential issue. It can be broken down as follows:
– The celerity of the flood wave (or propagation speed) in the watercourse, which determines the time available to obtain a relatively reliable forecast. This time makes it possible to decide when to move everyone in the protected area to safety.
– The flood rise gradient in the unprotected floodplain up to the spillway level, which affects the time required to take action before flooding occurs. This applies to zone A in particular (see below "Emergency management and levee managers").
– The flood rise gradient between the spillway level and the safety flood level. This will indicate how much time is left before potentially stronger flooding occurs. These two gradients are often assessed as the same value.

The celerity of the flood wave and the rise gradient(s) should not be confused, although they are strongly correlated. Celerity is associated with the intensity of rainfall, the valley slope, and the size of the upstream flood expansion areas (natural or controlled). The rise gradient may also be influenced by local conditions: the valley shape, flashy tributary inflows, etc.

When a flood with a given gradient enters the protected area, the most important parameter is the time between the beginning of overflow and flooding of a specific spot. In this case, we need to look at the celerity of the flood wave in the protected area. A 2D hydraulic model will allow us to assess this parameter. Outputs typically include a map of the water level (Figure 6.1), a map of the wave inflow time (Figure 6.2), a map of water velocity, and a map of the submersion time.

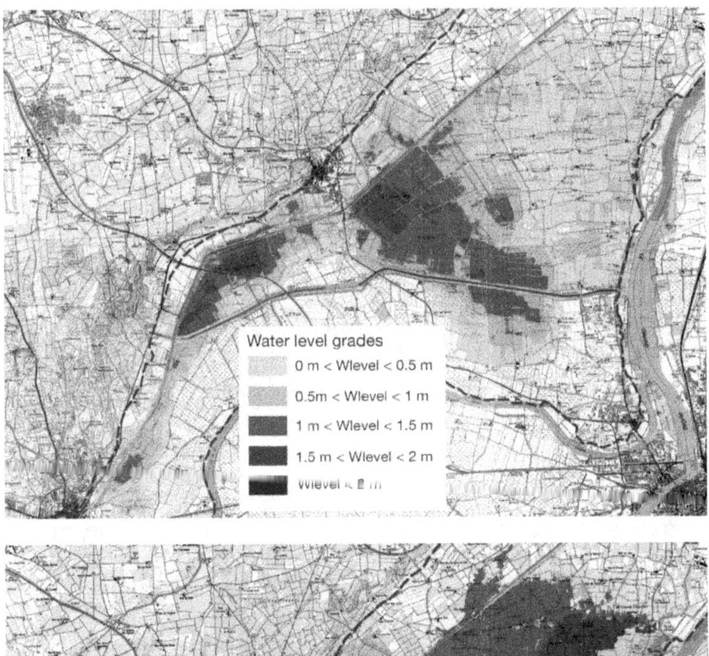

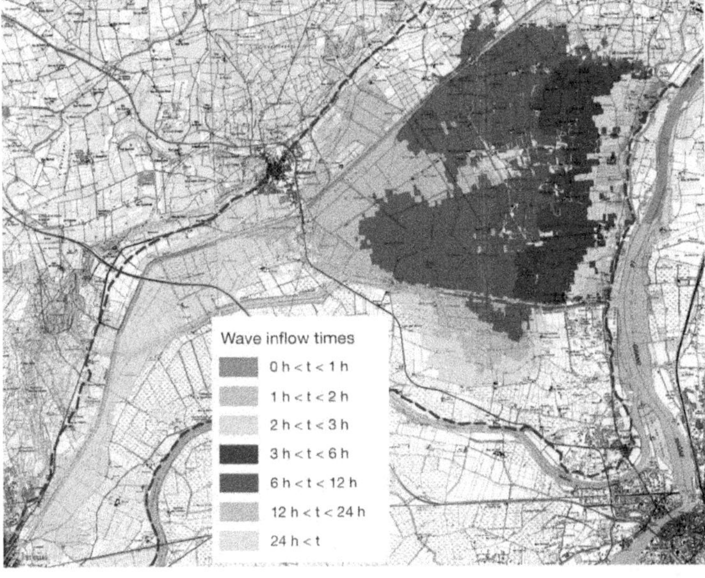

Figures 6.1 and **6.2.** Useful hydraulic parameters for emergency management (example of a typical 1856 flood that is between the protection level and safety level in a spillway project on the right bank of the Rhône River). (Source: SYMADREM, study to reinforce the levee between Beaucaire and Fourques carried out by ISL using the Rubar20 model)

Emergency management and levee managers

Levee managers must monitor the structures they are responsible for and, if needed, help ensure they are in working order. During the duration of the flood, they should also pass on information to any authorities responsible for ensuring people's safety (in France, the mayor and/or prefect). See below.

Flood zoning within a protected area

Given the considerations for protected areas and specific floods to watch out for, identifying the following zones could be useful:
– Zone A: where people must move to safety before overflows begin (this only applies if there is a safety spillway); for example, an area that will be reached by overflow in less than three hours.
– Zone B: where people should be moved to safety once overflows begin and before the safety level is reached.
– Zone C: where people should be moved to safety once the safety level has been reached.

These zones should be determined using a 2D hydraulic model, with additional safety margins. Splitting an area into several zones is only useful if there is enough time for people to move from one zone to the next. In contrast, there is little reason to create zones in a very small protected area or a flat protected area that fills like a basin. No such zones were established for the spillway projects in Aramon in 2003 and Comps in 2006 in response to rapid flooding of the Gardon River since the protected areas were small. On the contrary, even more zones could be created if the area is quite large and opens downstream (see Figures 6.1 and 6.2 above).

In France, mayors are responsible for this zoning as part of a Local Emergency Action Plan (LEAP) (see Section 6.5). They can draft this plan with the levee manager. Levee managers have no regulatory obligation to create these zones, but we think they are the most qualified to do so, particularly if there are several towns involved or in the case of joint local authorities (*syndicat intercommunal*), which are quite frequent in France. As a last resort, mayors are free to adapt the zoning in their LEAP.

The stages of emergency management

Any type of flood may potentially endanger people who live or work in the flooded area. If a levee fails, these risks are even higher. An emergency management strategy must therefore be developed, with saving lives as the top priority. Emergency management cannot be improvised and must be carefully thought through. It is important to understand, test, and assess all necessary measures to get people threatened by flooding to safety, particularly in the case of levee failure.

These measures should include ways to evacuate and shelter people according to the expected flood level, which is facilitated by the spillway. Safety measures may include preventively evacuating people outside the protected area, to higher ground in the area, or to safe refuges above flood level.

These measures should also include ways to inform residents about the risks and the measures that will be taken.

A crisis is a rather long period that may be broken down into four phases:
– Pre-crisis: The time between the alert and the feared event. It could be quite short and leave no time to prepare for the event. Or it could be quite long, allowing ample time to prepare. In any case, measures should be carefully planned beforehand. The pre-alert time is crucial since a crisis is rarely managed well if the pre-alert has not been given, meaning the response to the crisis was improvised. However, this issue is not specific to spillways.
– The feared event occurs: The peak of the crisis. It can be quite short (a few hours) or can last several days; this is when the danger occurs, as do rescue efforts. It is also a time for monitoring and emergency rescue measures.
– The return to normal: This could last from several days to more than a year, depending on when the effects of the event recede; this may also include draining of the flooded area, the victims returning home, a factory restarting, and so on.
– The implementation of measures to prepare for the next emergency or crisis: This period is generally very long (several years or even decades) and is often neglected. It includes additional or corrective measures based on lessons learned from past events.

For example, measures taken during Comoy's work programme in 1867 after disastrous flooding of the Loire River in 1856 and 1866 ended in 1891 when the last spillway was built in La-Chapelle-aux-Naux. The programme was left unfinished until 1907 when the final decision was made not to build a spillway in Givry – 41 years after the 1866 flood.

Preparedness is required to properly manage an emergency. The situation will never turn out as expected, but forethought and measures (resources, organisations, first response checklists, etc.) designed to respond to a potential emergency will always be useful when managing the actual crisis.

For example, on 1 January 2000, the Y2K bug did not have the expected effect in the Charente-Maritime department as the power was out in almost the entire area! On 27–28 December 1999, Cyclone Martin had destroyed a great deal of infrastructure, including coastal levees, and the resulting coastal flood killed several people. Though this crisis was as sudden as it was unexpected, the response was facilitated by the resources and organisation implemented through the "Polmar Plan", created after the sinking of the oil tanker *Erika* on 12 December 1999, during a heavy storm.

Though the 1999 event, called the *Storm of the Century*, damaged levees and caused many casualties, it did not receive as much press coverage as Cyclone

Xynthia ten years later. Although there was a similar number of fatalities in the department, this 1999 storm provoked so much other damage, such as fallen trees, that damage caused by levee failure was overshadowed.

During Xynthia, on 27 February 2010, local elected officials and firefighters evacuated the part of the town of Saint-Clément-des-Baleines that was exposed. This town had previously conducted a preventive evacuation in November 2002 after a breach in the large Noleaux levee. Although the 2002 evacuation proved unnecessary, it served as a practice run for local authorities and may have saved lives in 2010.

Spillway management during an emergency

Levee managers should pay close attention to spillways during an event that is likely to activate them.

Managers should warn relevant authorities (in France, the prefect, affected mayors, etc.) if they think the spillway will be activated by the expected flood and should warn the same authorities once the spillway begins operating. In addition, where a flood forecasting service (*service de prévision des crues* – SPC) exists, it should be notified if a substantial part of the watercourse's flow will soon be diverted.

Since water management will depend on the spillway's serviceability, managers should inspect the spillway before discharge begins and continue to monitor the spillway throughout its operation. This is even more necessary if the structure has a fuse plug or movable devices. To do so, managers must have identified monitoring access points during flooding and lighting devices must be ready to use. Monitoring staff must understand how the spillway operates and know what to look for in particular (floating debris that could create problems, erosion, etc.).

Spillway management during a flood must be an important part of the written instructions for the levee system. In France, these written instructions should be shared with the prefect of the department (the department that monitors hydraulic works).

During flooding, levee managers should make sure that all monitoring staff remain safe. It is also extremely important that they always have a clear path to safety if the water level increases above the level of the structures, and that they can be reached at all times.

Besides monitoring, it could be useful to plan ahead for any work that might need to be done on the spillway during flooding:
– Removing any floating debris (trees, caravans, etc.) that might limit the spillway's conveyance or cause damage as they pass over the spillway.
– Placing riprap or big bags of sand to prevent erosion from starting.
– Checking the alternative emergency operation of a movable device if any.
– Observing the behaviour of a fuse plug, etc.

To do so, all the necessary equipment and materials must be on hand or a contract should be negotiated with a contractor beforehand. These types of interventions can only be carried out under the supervision of engineers approved by the levee manager. Precautions must be taken to ensure their safety: off-grid communications in case the network gets saturated, emergency evacuation options if the water level rises, etc. Access paths must be identified and clearly visible.

The emergency management plan drafted by levee managers must describe each person's responsibility, particularly for levee technicians and any municipal technicians that may be involved. Proper communication between professionals (levee guards and engineers) and support staff is important to avoid endangering those who do not fully understand the risk involved. Interventions on the structure must exclusively be performed by levee managers or companies they commission.

The spillway should also be checked soon after overflow ends. If needed, plans should be made to repair the spillway to ensure it becomes safely operational as soon as possible, particularly if it has a fuse plug. It could also be useful to remove any alluvium deposits that have accumulated (see Section 4.4.2).

As an example, SYMADREM (Syndicat mixte d'aménagement des digues du Rhône et de la mer – Public establishment responsible for the management of river and sea levees in the Rhône delta) developed the following response procedure: each levee segment has four alert levels (from 0: pre-alert to 3: reinforced alert). At level 2, monitoring teams are on call from 9 AM to 5 PM. At level 3, 24-hour monitoring begins. As soon as issues arise, the on-duty staff notify the command unit, which in turn orders the local levee guard to go on-site. A joint decision is then made whether to bring in a public works company to intervene. Monitoring teams comprise SYMADREM engineers and any available staff or volunteers from the local communities.

Local Emergency Action Plan (LEAP)

Article 13 of French Law 2004-811 of 13 August 2004 on the modernisation of civil protection requires all municipalities with a Risk Prevention Plan (RPP) to establish as LEAP. This local plan, as described in Decree 2005-1156 of 13 September 2005, is integrated into the emergency response at a wider geographical level. Even for municipalities without an RPP, a LEAP is strongly recommended.

The LEAP "determines, based on known hazards, immediate preparedness and measures to ensure human safety, defines all necessary arrangements for disseminating alerts and safety instructions, identifies all available resources, and establishes the organisation of measures to support and assist the population".

The LEAP is established at the municipal level, but if the protected area includes more than one municipality, an **inter-municipal** approach is recommended. The previously mentioned decree (Article 5) provides that "municipalities that are members of a public establishment for inter-municipal cooperation with its own tax system may charge this entity with creating the inter-municipal emergency action plan and managing and obtaining any resources needed to carry out this plan". Otherwise, all local action plans should be coordinated; in this case, it would be helpful to have a single consultant responsible for all these plans.

The LEAP should be ready to put into action when it is needed. The goal is not simply to draft a document but to prepare the municipality to effectively respond to an emergency. The LEAP for a municipality in an area with a spillway should provide all the practical and necessary information about the spillway, including any instructions local inhabitants should follow if the spillway overflows. The LEAP explains what to do and what not to do.

The LEAP must indicate the moment when people should be moved to safety, almost always before the flood reaches the spillway. One example is the spillway in Aramon (Gard department) during the 2003 flood. The mayor issued an evacuation order when the water reached 30 cm below the spillway level. This was sooner than the plan required, but night was falling and a daytime evacuation was easier to manage.

The LEAP should detail all the tasks to be completed (sector-by-sector monitoring, information to residents, potential evacuation assistance, etc.) and the means of cooperating with the levee manager. Each person's role must be clearly defined, particularly if any community response workers become involved in crisis management.[12] Community response teams comprise volunteers who can support municipal staff in specific emergencies.

A LEAP may plan for municipal and community response teams to help the levee manager monitor the levee system, particularly the spillway. This is how SYMADREM and its member communities proceed (see above).

Over time, memory and experience tend to fade. Regular training sessions are required for the levee manager and local authorities to become used to working together. These sessions offer an opportunity to consistently update all "first response" checklists, which are critical in emergencies.

Special Emergency Management Plan for Flooding

In France, the prefect of each department must have an emergency response plan. It may be broken down into several specific plans. If the department is likely to experience flooding with severe consequences, it will include a Special Emeregency Management Plan for Flooding.

---

12. The French 12 August 2005 Interior Circular on community response workers; new Articles L. 1424-8-1 to L. 1424-8-8 of the Local and Regional Community Code resulting from the Law of 13 August 2004 on the modernisation of civil protection.

This last plan should identify the department's resources and the organisation of its response to flooding that is too extensive for local communities to handle on their own.

If the department does not have the resources to manage a large flood, the prefect may request assistance at the zonal level (zone of defence).

The Special Emergency Management Plan for Flooding should also describe how to ensure that people who are in danger behind a levee or a spillway get to a safe location.

7

Economic aspects

Gérard Degoutte

Economic studies are not generally carried out for spillways in isolation from the rest of the structure. The entire levee system has to be considered rather than a single component. This makes sense for a new levee project that includes a spillway. If a developer decides to install a spillway on an existing levee, some work will necessarily be done on the levee (reinforcement, adjustment of the profile) so the economic study will cover the entire levee refurbishment. Even if the project is only "adding" a spillway, the economic study will still consider the levee system as a whole.

An economic study for a flood protection project does not specifically address the presence of a spillway on a levee. A range of economic methods can be used, but we will offer a brief overview of the cost-benefit analysis (CBA). This entails:
– Estimating the structure's lifetime *(n)*. In the absence of national recommendations, we suggest using a lifetime of 100 years, and at least 50 years. However, the lifetime of a structure may evolve depending on changes to land use conditions during this time, which is always unpredictable. The chosen lifetime will therefore be a conventional one and a standard choice makes it easier to compare projects.
– Estimating investment expenses I and maintenance costs C for year i (which are typically consistent, taken as a mean year-on-year value).
– Estimating annual benefits B (or avoided damages, namely the difference between damages without the levee system and residual damages with it). These are direct and indirect damages (operating losses, etc.). The average annual benefit in statistical terms is

$$B = \int (D(f) - D'(f))df$$

(yellow-shaded area in Figure 7.1). Existing damage curves are available for property damage that can be expressed in monetary terms, excluding intangible damage (costs related to human casualties, accidents, psychological effects, etc.). We can also consider that protecting human lives is more dependent on an effective alert and evacuation system than the type of protective structure. This calculation assumes that we can estimate the damage before and after construction based on hydraulic parameters associated with how frequently the flood event in question

will occur (water level, overtopping time, velocities) within a wide range of events. This is rather challenging as we only have orders of magnitude to work with.
– Estimating the net present value (NPV)

$$NPV = \sum_{i=1}^{n} [(B - C)/(1 + r)^{i}] - I$$

where r is the defined discount rate.

In 2006, the Commissariat Général au Plan[13] (French Economic Advisory Committee) recommended applying a consistent 4% discount rate for the first 30 years, then a declining rate that would reach a 2% floor rate within 500 years. This decreasing rate can be obtained using the following approximate calculation:

$$r = \sqrt[n]{1.04^{30} \cdot 1.02^{n-30}} - 1 .$$

For example, the rate will be 3.2% at 50 years and 2.6% at 100 years.
– The project will be deemed economically viable if the NPV is positive.

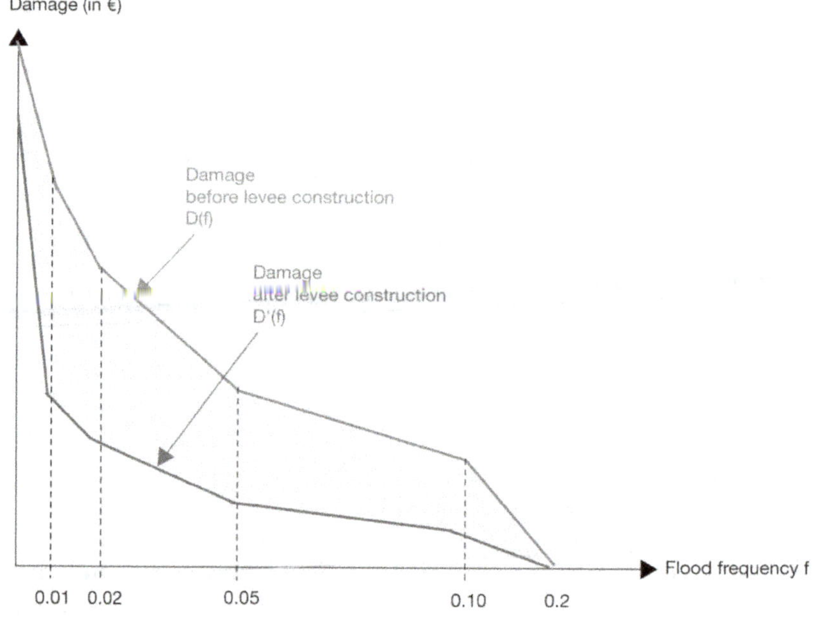

Figure 7.1. Example of a damage curve. This example shows how floods below the 5-year return period should not cause any damage, even before work is done.

We recommend two publications that cover this type of study since they provide valuable information about damage functions and present other economic methods besides CBA:

---

13. Report submitted to the Commissariat Général au Plan: "Révision du taux d'actualisation des investissements publics" ("Revised discount rate for government investments") on 21 January 2005; Expert panel chaired by Daniel Lebègue.

– Ministry of Ecology, *Évaluations socio-économiques des instruments de prévention des inondations* (*Social and financial assessment of flood prevention tools*). Study by the Department of Economic Studies and Environmental Assessment (D4E), based on a study carried out by Ledoux Consultants, CEREVE, and CEMAGREF in March 2007, 2007, 117 pages.

– The European Center for Flood Risk Prevention (CEPRI), *Évaluation de la pertinence des mesures de gestion du risque inondation; manuel des pratiques existantes* (*Flood hazard management measure assessment: existing practices manual*), June 2008, 36 pages + Appendices.

Both are available (in French) on the CEPRI website:

http://www.cepri.net/publications-et-documents.html

Some more recent documents (In French) are also downloadable at: https://www.ecologie.gouv.fr/levaluation-economique-des-projets-gestion-des-risques-naturels

English reader will find a summary of the French socioeconomic evaluation of flood prevention projects here:

https://www.ecologie.gouv.fr/sites/default/files/Théma%20-%20The%20socio-economic%20evaluation%20of%20flood%20prevention%20projects%20in%20France.pdf

8

Summary and Conclusion

Gérard Degoutte

The benefits and limits of a spillway in a protected area

This is only a question in protected areas since a spillway is always necessary for a controlled flood expansion area (other than in rare situations in which pumps or siphons supply water into the area).

Installing a spillway in a protected area offers many benefits:
– The spillway will delay overflow on other parts of the levee (generally to a limited extent).
– The spillway can introduce a stilling water cushion before water overflows on other sections of the levee, except in specific configurations with steeply perched valleys.
– If this water cushion is well formed, it reduces the need for additional structures to make the levee overflow resistant.
– However, even if none of these expensive structures are built, and overtopping causes a breach, it will be less violent thanks to the downstream water cushion and will occur after the area has been evacuated.
– In any event, the spillway will reduce the risk of a levee breach and therefore substantially reduce the discharge and the volume of water. It also reduces the water levels, the velocities, and the flooding time in the protected area.
– The location of the flows can be chosen ahead of time, whereas the location of a breach will remain unpredictable.
– The moment overflow begins can be accurately predicted, and the discharge law is well known, facilitating emergency management (this benefit is not as significant with a fuse plug).

Despite offering all these benefits, a spillway does not eliminate the risk of levee failure, unless the entire levee has been made overflow-resistant or unless the water level in the protected area reaches the levee crest before overtopping begins.

Preparedness for emergency management remains essential and the spillways' presence must be included in all urban planning documents.

Spillway features

This section will cover both protected areas and flood expansion areas unless stated otherwise.

Hydraulics

The lateral spillway flow law is more complex than for a frontal river weir due to the distribution of the approach velocities. For large projects, it would be wise to develop a scale model. If the spillway is not directly connected to the river-bed but the unprotected floodplain, the velocities of approach should be closely monitored as they might change according to the topography and vegetation in the unprotected floodplain. They could also be affected by constructions if we do not have control over land use!

In flood expansion areas, it is important to set the spillway length and the moment when the flow diversion begins. A flood expansion area should not fill with water too quickly or it will be less effective. In other words, a flood expansion area should not be used for low discharges that would not cause any harm farther downstream. On the contrary, it should be used for the first floods that are harmful since no one would accept damage downstream of the spillway while the flood expansion area remains dry! Proper calibration is therefore fundamental and must be done after testing many floods, particularly using different hydrograph rise gradients.

Civil engineering

The most important aspect of river levee spillways is that they are typically very long structures that operate with low hydraulic loads if they do not feature a fuse or movable device. They will therefore only experience moderate loading, allowing for a different civil engineering design than spillways on dams. In particular, costly constructions made of reinforced concrete with construction joints every 12 m or 15 m are unnecessary. Solutions using materials such as concrete riprap or Reno mattresses are flexible and can be adapted to traditional levee configurations. Yet these structures are not simple. Advanced engineering is required to ensure that the spillway, which is meant to offer greater protection, does not become the weakest spot. This requires close monitoring of the risk of internal erosion at the contact point between the embankment and the spillway's sidewalls or apron, etc. The structure should be carefully designed with anti-piping shields, filters, drain outlets or weep holes, properly compacted soil under (the "foundation" component) and against the spillway if there are vertical sidewalls, and so on.

However, good practices are currently emerging in this field, and we hope this guide will contribute to this development.

Several techniques already used on dams could be considered for future engineering projects, such as Hydroplus fusegates or piano key weirs that cut the length needed in three.

En piste d'avenir, on retrouve des techniques issues du monde des barrages, par exemple les seuils en touche de piano qui peuvent être trois fois moins encombrants en longueur, ou les hausses Hydroplus.

Site Selection

Site selection cannot be based on a single consideration, like the idea that a downstream site will reduce velocities in the protected area or the flood expansion area; or that an upstream site will improve flood attenuation by using the flood expansion area more efficiently. Other criteria must also be considered, such as proper filling of the landside levee toe, concave banks, the presence of an unprotected floodplain, etc.

Simulations will often be needed to support the final decision, rather than relying on first impressions, no matter how expert they may be.

Geomorphology and sediment transport

The spillway itself can have morphological effects by promoting riverbed aggradation. This is not a big problem if the spillways start diverting water for floods that are rare enough, which is highly recommended in terms of hydraulics in any case.

Each time the spillway goes into effect, diffluent currents might cause deposits to settle near the spillway's downstream end. The deposits could be disruptive the next time the spillway is activated, especially if they are fixed by vegetation. This can be prevented through routine maintenance, which should be included in the flood instructions.

More worrisome is the case of watercourses that are not in dynamic equilibrium whether they are aggrading or degrading. In this case, the spillway will be activated too early or too late. There are solutions to this problem, but a comprehensive geomorphological analysis is required.

All these factors show the need for regular operational reassessments. This could easily be done as part of the risk assessments that need to be updated every 10 years in France for the larger levee systems, and at longer intervals for the others.

Operation

Given that spillways must function effectively during the flood event, it is important to install robust structures that do not require manual intervention, if possible. However, movable devices may also be used as long as they are sturdy, the levee manager has the technical skills to monitor and maintain them, and they are well accepted by local inhabitants (in protected areas).

Monitoring and maintenance

Trapezoidal weirs are preferable to rectangular weirs to facilitate the movement of vehicles on the levee crest outside of flood events.

Monitoring is critical to ensuring the spillway will remain operational based on the assumption that a flood event is always imminent, all year long.

Feedback on spillway operation during floods is valuable and should benefit the entire working community, particularly as part of ICOLD (International Commission on Large Dams), which created a Technical Committee on levees in 2017 that officially extended its work to coastal and river flood protection levees.

Emergency Management

Spillway operation in a protected area will likely prevent very serious crises if the spillway is well designed and the levee is in good condition. Yet the presence of the spillway requires emergency management since it will flood the protected area. It is important to emphasise the need to raise awareness among residents in the protected area and conduct alert training exercises. Yet this would also be important even in the absence of a spillway.

Is a spillway always required in a protected area?

This question only applies to protected areas since a spillway is always needed to feed a flood expansion area with levees.

Nor is this a question when significant flow volume can be diverted to a flood-water retention area (like the Lironde River) or an area with low stakes (like the old and new spillways on the Vidourle River). These are ideal situations for building a spillway.

As a result, the question about the necessity of a spillway only applies to levees in protected areas.

We have shown the benefits of these spillways throughout this guide and summarised them in the beginning of the chapter. We have concluded that spillways are frequently useful in new or existing protected areas. However, there are several configurations in which a spillway is of limited use, or may even be inappropriate.

Levees that are evenly overflow resistant

Some levees in protected areas may be overflow resistant across the entire profile. They are made of rigid materials, such as concrete or masonry, and must be properly designed with good foundations. They could also be made of armoured earth and be extended by an apron or a stilling basin. It is important to remember that a levee's weakest spot is the downstream toe (see Chapter 1). With these levees, the spillway cannot play its protective role since the levees are not likely to fail, even if they overtop. Nevertheless, a spillway could still serve a purpose by filling the area with water in a controlled way and facilitating the warning process. For evenly overflow-resistant levees, a spillway seems neither essential nor useless.

However, for a spillway to be truly unnecessary on an overflow-resistant levee, every part of the levee must be resistant to overflow. For instance, on the Vidourle River, a 1 km section on the left bank was made overflow resistant in 2004 thanks to large rockfill concrete armouring (Photo 1.5). But the Pitot spillways were maintained since their cumulative conveyance minimises the risk of overflow on the rest of the left bank levee farther downstream. The spillways also help protect the levee on the opposite bank.

Rivers with high sediment transport

For rivers with very high bed load transport, the spillway may be at risk if the river is not in dynamic equilibrium. Otherwise, it may be challenging, costly, and ecologically unsound to attempt to maintain the riverbed's longitudinal profile. There is also the risk that the spillways will be activated too early if the riverbed is aggrading, or too late if it is degrading. Or the spillways may never be activated if the longitudinal profile is tilting, thereby causing the straight section of the levee to overflow while the spillway remains unused.

Torrents

Most torrents are leveed on their alluvial fans (e.g., the Doménon River mentioned in Chapter 4 and Photos 4.2 and 4.3). A spillway will not necessarily serve a hydraulic purpose since the waterline is not lowered upstream and no water cushion is created on the side of the alluvial fans. Most importantly, during flood events when the spillway is meant to operate, heavy bed load, wood jams, or even debris flow can occur and hinder the spillway's serviceability or even make it harmful.

> ### Conclusion on the need for a spillway
>
> In the case of rivers, we do recommend installing one or more spillways as a basic solution for a leveed protected area. However, the feasibility and utility of this spillway must be carefully reviewed, and the decision not to use a spillway must be justified.
>
> On the contrary, in the case of torrents, we recommend not using a spillway unless its presence can be justified.

Concluding remarks

Spillways on levees play the same role as those on dams: they convey water during floods to protect the structure they are associated with.

However, spillways convey water very differently on each of these structures: on dams, they evacuate excess water that needs to be removed into the river downstream. On levees, the water is not released but held for a given time span. It will either be temporarily stored in flood expansion areas to attenuate flooding, or temporarily stored in protected areas to protect the downstream slope of the levee and diminish flows.

Though it goes without saying, it is still worth mentioning: adding a spillway to an old and poorly understood levee is like bandaging a wooden leg. If the spillway is old as well, it is essential to determine whether the levee will fail before water reaches the spillway (if the safety level is inadvertently lower than the protection level – which would then just be an apparent protection level).

Levee spillways are generally much longer than dam spillways and tend to experience much lower hydraulic loads. As a result, they are usually lightweight and flexible structures requiring less robust construction. Promising best practices are currently emerging, though we still have limited feedback on spillway performance.

On a dam, the taller the spillway, the better. However, a levee spillway requires a more refined hydraulic design: neither too low, too high, too short, nor too long.

We therefore encourage engineering firms, and most importantly the owners and managers of the structures, to perform comprehensive studies. It would be easy enough to build well-designed structures and just make sure they improve the situation. We would simply choose an appropriate site and arbitrarily set the design protection level. But it would be even better to optimise the design to improve hydraulic efficiency. A good levee spillway project requires sustained trial and error with many simulations. Hydraulic tools are available, and can even be used to present projects to decision-makers and local residents in clear and understandable terms.

Spillways in flood expansion areas serve to attenuate strong floods. They partially compensate for land-use planning choices that ventured too far into floodplains. They could also be used to offset environmental losses by introducing water every year to restore wetlands. This use is beyond the scope of this guide, but an integrated project could include a spillway designed to protect human populations and a siphon to re-create the original semi-terrestrial/semi-aquatic environments. We recommend conducting an ecological study on principle, which would help determine whether these wetlands can be re-created.

Spillways in protected areas are designed to protect levees so they are less of a threat to the communities they are meant to protect. These should be associated with urban planning provisions to avoid creating a lose-lose situation: increased protection reduces potential hazards, increasing the likelihood that more people and assets will be introduced, and therefore creating more risk.

Lastly, no good guide can conclude without suggesting future avenues for research! One idea is to look for techniques that would make the soil resistant to overflow. We mentioned lime treatment to substantially improve resistance to erosion with no environmental impact. There is also a need to better predict the level of erosion control offered by a stilling water cushion. Since documentation of spillway performance is still rather limited, it would be worthwhile for a countrywide or international project to investigate structures that have already been activated, including both levee spillways and flood storage reservoirs. It is also important to monitor and examine future spillways constructions to better understand and improve resistance to external erosion.

Appendix 1

Spillway types on river levees, according to their hydraulic function and their purpose in flood management

Rémy Tourment and Adrien Rullière

River levee spillway: All spillways are designed to allow water to flow over them without damage when the water level is above their crest. This is a local function involving hydraulics and structural resistance.

The following functions of river levee spillways are related to their role in the system (Source-Pathway-Receptor, or waterside / protection system / landside) in terms of hydraulics.

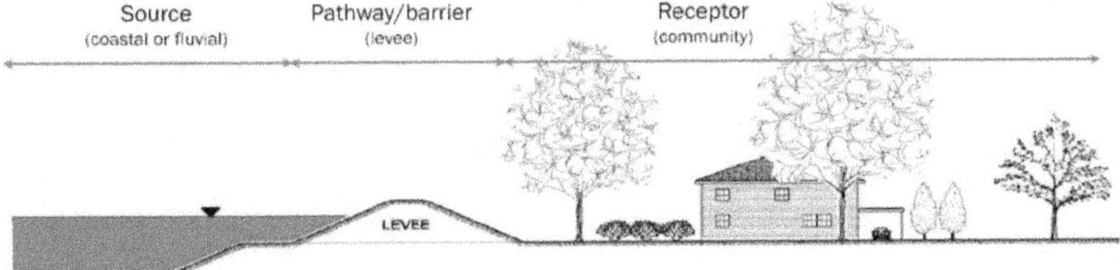

Figure A1.1. Cross-section of a levee.

Safety spillway: A spillway whose primary function is to avoid a breach in the levee system it belongs to by preventing overflow over a non-resistant levee segment.

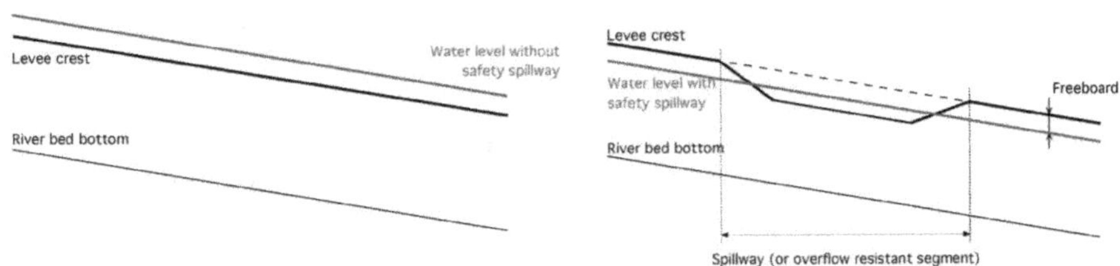

Figure A1.2. Safety spillway.

Diversion spillway or Diverter spillway or Flow divider spillway: A spillway that diverts part of the river's flow into the area protected by the levee system. Most river levee spillways are diversion spillways, including safety spillways and bypass spillways.

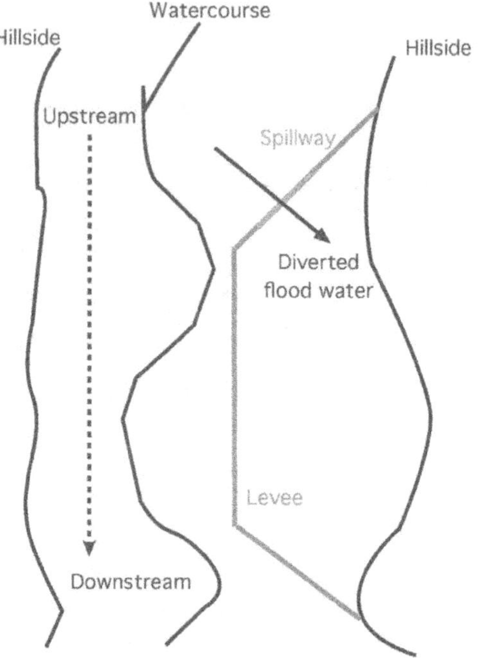

Figure A1.3. Diversion spillway.

Diversion weir or flow divider weir: When associated with a diversion spillway on a levee system, a weir in the riverbed can help control the proportion of flow that either remains in the riverbed or is diverted to the floodplain behind the levee.

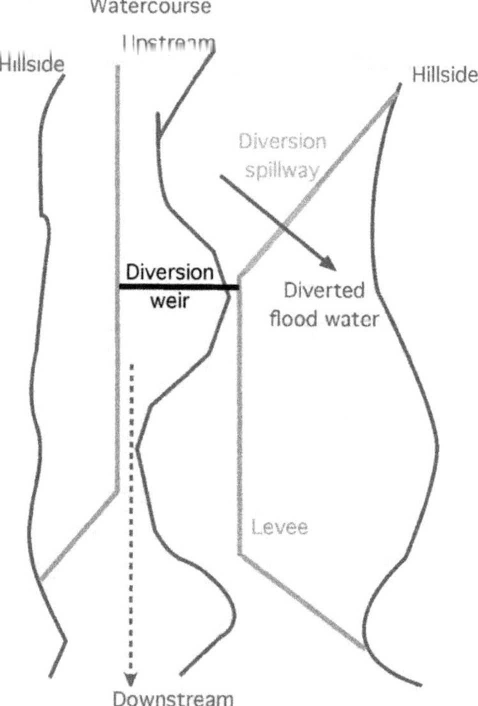

Figure A1.4. Diversion weir.

Return spillway: A spillway that lets water flow from the leveed area back to the river to avoid a breach when the water level is higher in the leveed area than in the river and higher than the levee crest. This is the case when the flood is receding quickly in the river, or if the incoming flood travels faster in the floodplain (from a spillway or a breach) than in the riverbed.

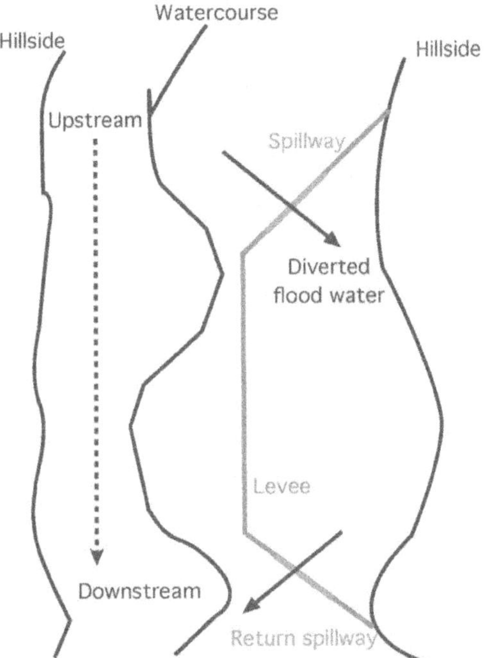

Figure A1.5. Return spillway.

Bypass spillway: A spillway that diverts part of the river's flow into a channel (natural or leveed) that goes back into the river downstream. This spillway and the associated channel bypass a segment of the river, usually because in this segment the channel is too narrow and/or because there are nearby people and assets to protect. A bypass spillway is a particular type of diversion spillway.

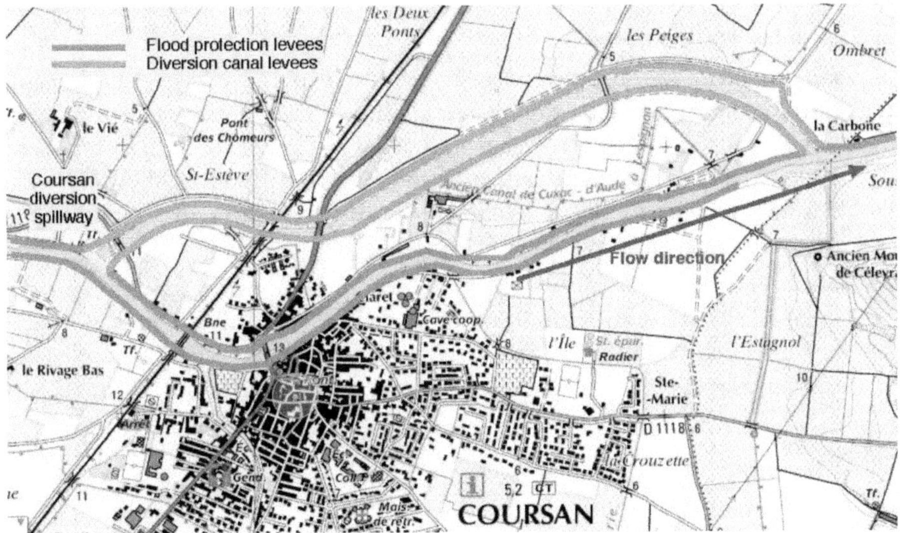

Figure A1.6. Bypass spillway.

The Coursan diversion spillway and canal, to the north of the Aude River (flows from left to right). See also Figure 1.10 La Bouillie bypass spillway.

Partitioning Spillway: A spillway that diverts part of the river's flow into a drainage channel leading to a different water body (lake, pond, sea, another river). See Figure 1.11: The diverter spillway built on the Lez River flows into a channel leading to another river that leads to a pond.

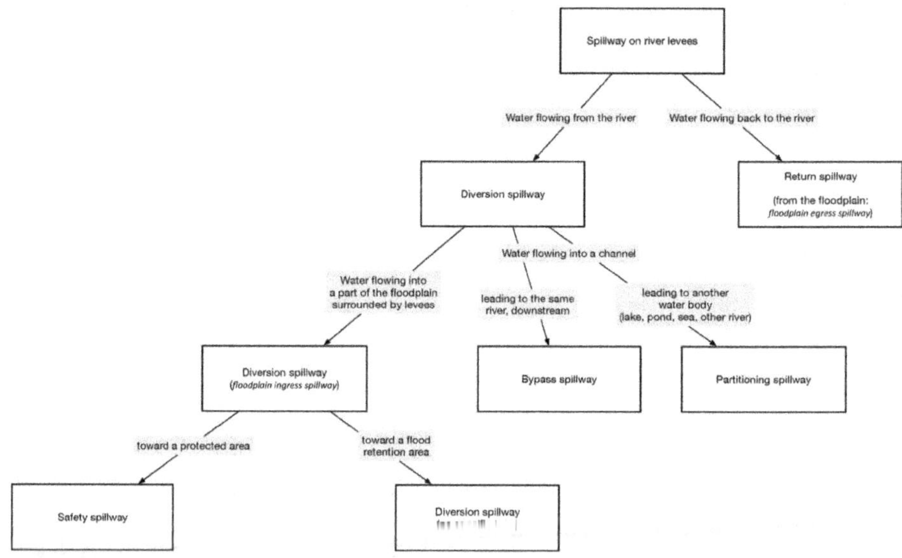

Figure A1.7. Flowchart of spillway types.

Variable spillway weirs:

Fuse plug spillway: A spillway that is meant to be removed by water. It is made of earth or gravel fuse plug ribbons that will be removed through erosion; a gravel ribbon topped by concrete slabs that tilt when water discharge erodes the ribbon; or Hydroplus fusegates that tilt under water pressure. The weir is lowered during a sufficient spill (in height and/or duration) and remains in that position (the weir is partially or totally lowered).

Movable spillway: A spillway with an adjustable movable device, flap gate, or inflatable weir (see the definition for this term), etc. This system can be adjusted and operated according to the water level. Contrary to fuse plug spillways, partial lowering is possible. After the spill, the weir can return to its initial position.

A history of spillways
on the Loire River

Jean Maurin

The oldest part of this historical review is based on the work of Roger Dion (1934).

The first spillways on the Loire River in France seem to date back to the end of the 16th century or the beginning of the 17th century. At the time they were called bypass spillways. Their function was to unload the Loire's flow upstream of a constriction in the leveed riverbed.

At the beginning of the 17th century, there was a bypass spillway in La Bouillie to protect the bridge in Blois and another one in Saint-Martin-sur-Ocre upstream of Gien.

In 1629, Louis XIII's Council recommended widespread spillway construction.

Following the 1707 flood, Louis XIV's government created a restoration plan in 1711. It called for filling in breaches opened by the last floods, raising the levees to 22 feet (about 7 m), enlarging them in proportion to the height, and building bypass spillways in areas that did not have any.

A construction programme between Gien and Tours led to the installation of several bypass spillways in the narrowest part of the leveed riverbed. We do not know their exact number or precise location, but archives provide us with information about some of them. In addition to the previously mentioned bypass spillways in La Bouillie and Saint-Martin-sur-Ocre, there were two in the Orléans leveed area near Sigloy, two in the Cisse leveed area near Négron, and one in the Tours leveed area between Montlouis and Tours.

The Pentecostal flood of 1733 overflowed the levees and caused many breaches. The flood destroyed the bypass spillways, creating the same damage as the breaches in the protected leveed areas. Engineers had doubts about maintaining the bypass spillways, so under the pressure from local residents, all of them were closed except the ones in La Bouillie and Saint-Martin-sur-Ocre (preserved at the request of engineer Louis de Règemorte). These two bypass spillways still stand.

In the 19th century, the three catastrophic floods of 1846, 1856, and 1866 caused more than 150 breaches. As a result, a Flood Commission was formed out of the

Conseil Général des Ponts et Chaussées (General Council for Civil Engineering), chaired by the General Engineer Guillaume Comoy. In 1867, Comoy called for the construction of 20 spillways to manage floodwater and allow for controlled flooding of the leveed areas. After studies and consultations with local residents, 11 spillways were authorised. Between 1867 and 1891, eight spillways were built. The last three were postponed and never built (Nevers, La Charité, Onzain).

The flood of 1907 revived fears and led the government to conduct an inventory in 1911. In 1917, two spillways that were not planned by Comoy were built on the Divatte levee, downstream of Angers. Then came World War I, and the government decided to abandon the projects in 1925.

There are now 15 spillways on the Loire River levees: the first two bypass spillways (La Bouillie and Saint-Martin-sur-Ocre), the eight spillways from the Comoy programme, the two spillways on the Divatte levee, and three at ground surface level. The latter are not civil engineering structures, but simple interruptions in the levees. The Passy "spillway" is in the middle of the Beffe leveed area, and the Bonny spillway is downstream of La Charité-sur-Loire. The Léré spillway allows water to flow between the hillside on the left bank and the nuclear power plant in Belleville-sur-Loire. The Mazan spillway allows flooding of the Ardoux leveed area.

While the spillways at ground surface level and the old bypass spillways have gone into effect many times, the spillways built during the Comoy programme have never functioned (except for the Bec d'Allier spillway, which overflowed in 1907).

During the 2003 flood, the Passy and Léré openings and the Saint-Martin-sur-Ocre spillway were activated. The Bec d'Allier spillway, which had functioned in 1907, had a freeboard of about 10 centimetres while the La Bouillie spillway had about 80 cm.

Table A2.1. The 15 spillways on the Loire River in 2012 (from upstream to downstream)

The spillways built during the Comoy programme are listed in red.

Name of the leveed area (and bank)	Name of the spillway	Characteristics	Time of construction
Bec d'Allier (left bank)	Guétin	Length: 400 m The crest is 4 m above low water Functioned in 1907	1870
Beffe-Bonny or La Charité (left bank)	Passy	Ground surface "spillway"	
Léré (left bank)	Léré (or la Madeleine)	Ground surface "spillway"	
Gien (left bank)	Saint-Martin-sur-Ocre	200-m-long bypass spillway without a fuse plug	End of the 16th century – beginning of the 17th century
Dampierre (right bank)	Pierrelaye	Length: 150 m	1867

Table A2.1. The 15 spillways on the Loire River in 2012 (from upstream to downstream) (continued)

Saint-Benoît (right bank)	Ouzouer	Length: 800 m Fixed weir at 5.3 m above low water level and topped with a 1 m fuse plug embankment	1886
Orléans (left bank)	Jargeau	Length: 575 m The crest is 5.5 m above low water level and topped with a 1.5-m fuse plug embankment The spillway was constructed on the 1856 breach location	1878-1882
Ardoux (left bank)	Mazan	Ground surface "spillway"	
Avaray (right bank)	Avaray	Length: 550 m Its characteristics are similar to the Ouzouer spillway	1883-1887
Upstream Blois (left bank)	Montlivault	Length: 400 m The crest is 5.3 m above low water level and topped with a 1.5-m fuse plug embankment	1887-1890
Downstream Blois (right bank)	La Bouillie	200 m long bypass spillway with a fuse plug embankment	End of the 16th century – beginning of the 17th century
Bréhémont [14](left bank)	La Chapelle-aux-Naux	200-m-long with a crest 5 m above low water level	1888-1891
Vieux-Cher[14] (left bank)	Vieux-Cher	100-m-long with a crest 4.8 m above low water level	1888-1891
La Divatte	Bel Air		1917
La Divatte	Petits Champs		1917

Table A2.2. The 20 spillways of the Comoy programme (1867)

Name of the leveed area and spillway	Characteristics	Time of construction
Saint-Eloy, upstream of Nevers	Never built	
Bec d'Allier leveed area, Guétin spillway	Length: 400 m The crest is 4 m above low water level	1870
Givry leveed area	Never built (in 1907 the chief engineer of the Nièvre department thought the spillway would cost more than the value of the properties to be protected)	

---

14. The two spillways in the Bréhémont leveed area were built at the location of breaches that occurred during the 1856 and 1866 floods.

Table A2.2. The 20 spillways of the Comoy programme (1867) (continued)

La Charité leveed area, Rauche levee	Never built due to disagreements about cost sharing	
Léré leveed area, Bannay levee	Never built	
Ousson leveed area, upstream of Briare	Never built	
Saint-Firmin leveed area, across from Briare	Never built – not needed following construction of the Briare water bridge	
Dampierre leveed area, Pierrelaye spillway	Length: 150 m	1867
Sully-sur-Loire leveed area	Project abandoned in 1874 due to local opposition	
Saint-Benoît leveed area, Ouzouer spillway	Length: 800 m The masonry fixed weir is 5.3 m above low water level and topped by a 1 m fuse plug embankment On the leveed area side, the spillway is supported by an 18-m-wide horizontal masonry berm, followed by a second 20 m rockfill berm	1886
Orléans leveed area, Jargeau spillway	Length: 575 m The crest is 5.5 m above low water level and topped by a 1.5-m fuse plug embankment The spillway was built at the location of the 1856 breach	1878-1882
Orléans leveed area, between Jargeau and Orléans	The project was abandoned as it was not expected to achieve the desired result	
Avaray leveed area, Avaray spillway	Length: 550 m The characteristics are similar to the Ouzouer spillway	1883-1887
Blois leveed area, Montlivault spillway	Length: 400 m The crest is 5.3 m above low water level and topped by a 1-m fuse plug embankment	1887- 1890
Ménars leveed area	Project dropped in 1874 – thought to serve no purpose	
Cisse leveed area, near Chouzy	This project, which would have required raising the railway tracks from Paris to Tours between Chouzy and the Montlouis Bridge, has been indefinitely postponed (1867–1938). The (railway) Compagnie d'Orléans refused to pay for half of the costs, as requested by the French government. The Indre-et-Loire General Council has protested this postponement several times.	
Luynes leveed area	Project dropped in 1869 due to local opposition	
La Chapelle-aux-Naux leveed area (on Bréhémont island)	200-m-long with a crest 5 m above low water level	1888-1891
Vieux-Cher leveed area	100-m-long with a crest 4.8 m above low water level	1888-1891
La-Chapelle-aux-Naux leveed area, Rupuanne	The project was abandoned as it was not expected to achieve the desired result	

Appendix 3

Overflow levees and spillways in the CNR's Rhône River development

Gilles Tratapel
Francis Fruchart

General principles

The levee systems on the Rhône River include non-overflow levees, meaning that they maintain a freeboard during a 1,000-year flood, overflow levees (which we refer to as overflow-resistant levees in the rest of the guide) and spillways.

When designing the Rhône River developments, CNR kept the main flood-prone areas in the Rhône Valley by turning them into flood retention areas.

During an overflowing flood, the Rhône's floodplain is inundated by one or more of the following causes (Figure A3.1):
– Run-off over a spillway.
– Backflow through a siphon.
– Flooding from downstream (backwater).
– Overflow directly into the floodplain.
– "Gentle" overtopping of the overflow levees with a water cushion already formed on the plain side.

This allows for gradual flooding of the plain. It can later be drained naturally or by gates, siphons, or pumping stations. The structure's operation, which is either active or passive depending on the situation and flow rate, determines its effectiveness.

Overflow levees function under a low hydraulic load and spillways under medium or high hydraulic loads. Overflow levees can operate over variable lengths depending on the floods, while spillways have a fixed length.

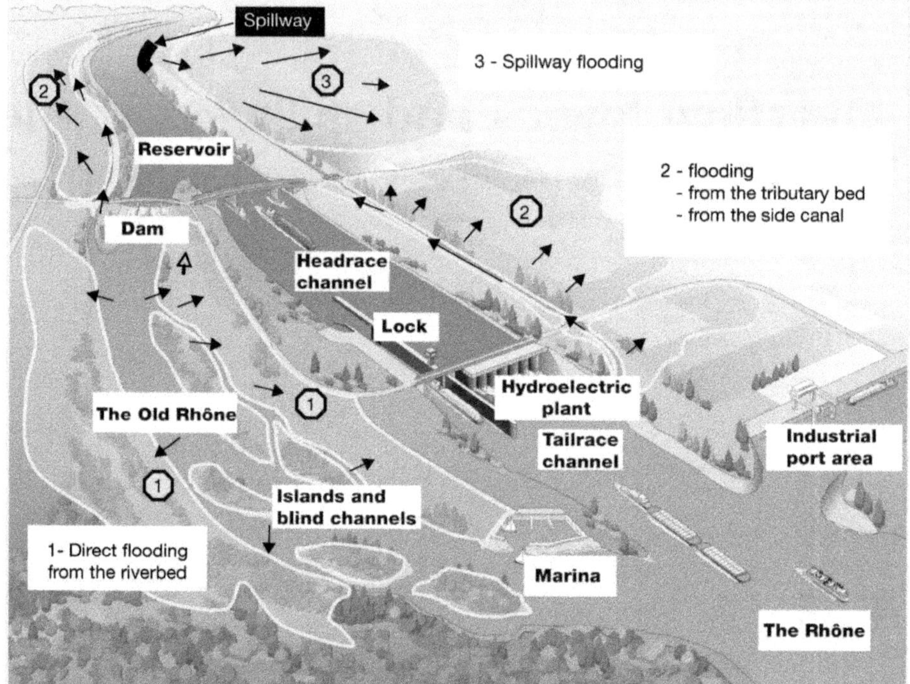

Figure A3.1 Diagram of a CNR development.

Spillways are earthen structures protected by either concrete rockfill, concrete slabs, or sheet piles. They can be equipped with floodgates.

An overflow levee can withstand overflow over variable lengths without damage. In general, there is already a downstream water cushion by the time the levee overtops. To ensure this cushion forms, the overflow levee is designed with a steeper upstream-downstream slope than the river's waterline. As long as the flow over the levee remains submerged, the downstream face does not require protection (the overflow is subcritical). When the levee is in free flow, which can occur on the upstream part of the overflow levee, rockfill protection is needed. It is better to avoid concrete facing, which is too smooth and transfers erosion to the levee toe.

Submersion conditions are defined in the specifications and the construction terms of reference, approved by the State:

– Non-overflow levees are designed to prevent overtopping for flows less than or equal to the 1,000-year flood and maintain a freeboard at this level. However, some areas behind these levees can be inundated for floods lower than or equal to the 1,000-year flood through tributaries or activation of a spillway.

– Overflow levees are designed to overflow at flows lower than the 1,000-year flood. They represent around 10% of all CNR levees and are primarily found on power plant impoundments (particularly on the upper Rhône) or in tailrace canals (the Saint-Vallier, Bourg-lès-Valence, and Vallabrègues developments).

Regarding the distinction between "wet" and "dry" levees, the former function much like dams (with small variations in the water level) whereas the latter only operate during major floods.

The following table comes from a comprehensive study of the Rhône River conducted by Territoire Rhône at the initiative of the DIREN[15] (Regional Environment Authority) for the Rhône catchment area. It summarises the discharge attenuation by the main flood expansion areas, for medium floods (around 10 years), large floods (around 100 years) and very large floods (around 1,000 years). We can see that there is a significant discharge reduction in large floods on the Rhône by cumulative effect (23%).

Table A3.1. Flood discharge reduction in m³/s and % of peak discharge upstream of the main floodplains.

Floodplain	Medium flood		Large flood		Very large flood	
	Net discharge reduction (m³/s)	% of peak discharge	Net discharge reduction (m³/s)	% of peak discharge	Net discharge reduction (m³/s)	% of peak discharge
Chautagen – Lac du Bourget	110	7	570	7	885	35
Lavours – Yenne plain	70	5	150	7	255	10
Bangues plain – Saint-Benoît	150	9	175	8	485	19
Miribel island – Jonage	60	1.5	70	1.6	320	6
Total upper Rhône	400	13	1000	23	2000	40
Livron plain (Printegarde)	25	0.5	20	0.3	30	0.3
Tricastin plain	210	3.5	200	2.6	960	9.5
Caderouuse-Codolet plain	0	0	90	1	70	0.5
Roquemaure plain – Oiselet and La Barthelasse islands	80	1	65	0.5	120	0.8
Aramon plain – Montfrin and Vallabrègues-Boulbon	20	0.2	160	1.3	160	1.1
Total lower Rhône	350	4	550	5	1400	10
TOTAL RHÔNE	**750**	**9**	**1550**	**13**	**3400**	**23**

---

15. Now known as DREAL.

These flood expansion areas can be fed by spillways (Vallabrègues, Caderousse, Livron, etc.), siphons (Lavours swamp), floodgate structures and overflow levees (Brangues-Le Bouchage-Saint-Benoît), upstream backwater (Aramon-Montfrin, Codolet), or natural overflowing of the Rhône River (Miribel-Jonage, Sablons, Donzère-Montdragon, Barthelasse island).

Below we offer two examples to illustrate the functioning of two flood expansion areas first supplied by spillways then by overflow levees. One is on the upper Rhône, in the Chautagne plain and Lac du Bourget area, and offers significant discharge reduction downstream (35% for a 1,000-year flood). The other is on the lower Rhône, in Printegarde (or Livron), and offers lower absolute discharge reduction, and much lower relative discharge reduction (0.3% for a 1,000-year flood).

Example 1: The Chautagne plain and Lac du Bourget

This vast plain on the left bank of the Rhône (approximately 3,000 ha) is a flood-prone area that the CNR development turned into a flood retention area. A spillway and overflow levee fill the Chautagne plain and the Lac du Bourget with floodwater, contributing to significant discharge reduction of Rhône flooding in the area since the peak discharge of the design flood (around a 1,000-year

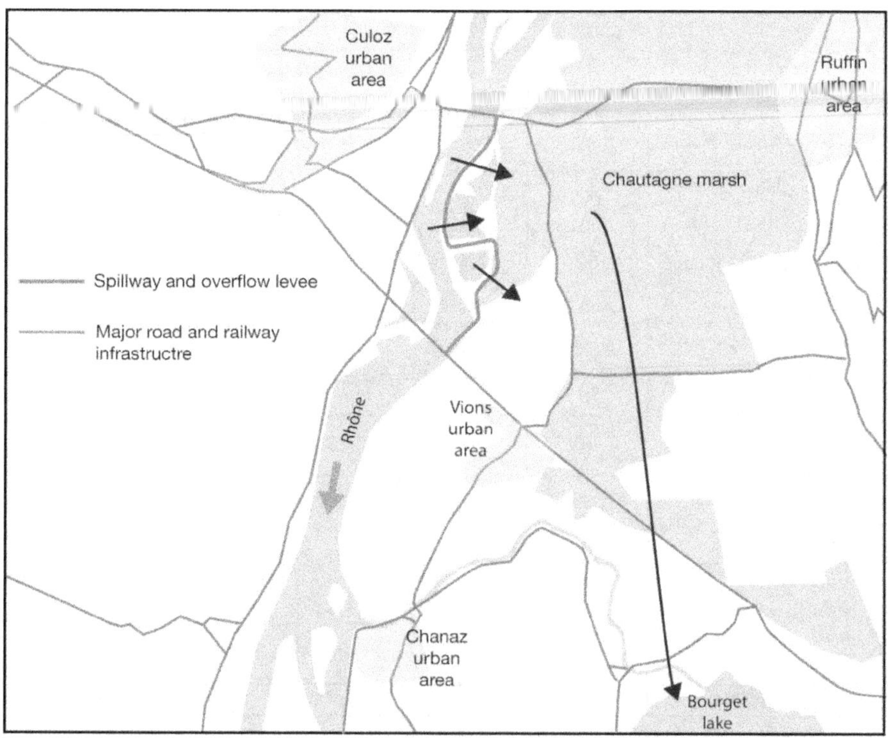

Figure A3.2. Chautagne flood expansion area, Lac de Bourget.

flood) decreases from 4,150 m³/s upstream to 2,800 m³/s downstream. This is why the upstream Motz dam (Chautagne development) has five floodgates, while the downstream Lavours dam (Belley development) only has four.

Upstream of the Belley dam, the levee on the left bank of the power plant impoundment overflows between the Loi Bridge (north) and Vions Bridge (south), as depicted in Figure A3.2. At the downstream end of this overflow levee, there is a spillway that is 60 cm lower than the levee crest. At the upstream end, the overflow levee is connected to a non-overflow levee that is 90 cm higher than the overflow levee's crest.

The spillway is activated by a flow rate of around 700 m³/s. A water cushion starts to form at the back and protects the overflow levee when it begins to over-top at the downstream end. As the flow continues to increase, the water cushion expands upstream, as does the overtopping of the levee.

Table A3.2. Flow rate when overflow begins for a 10-year and a 100-year flood

	Overflow levee			Spillway
	Upstream end (starting 0 km)	Middle (0.55 km)	Downstream end (1.55 km)	1.7 km
10-year flood	No overflow	1,608 m³/s	1,413 m³/s	738 m³/s
100-year flood	1,976 m³/s	1,581 m³/s	1,326 m³/s	721 m³/s

This table clearly shows that overtopping of the overflow levee progresses from downstream to upstream and spills onto a water cushion that also moves upstream. The spillway begins operating for a 10-year flood 24 hours before the flood's peak discharge and between nine and 12 hours before a 100-year flood. The overflow levee's downstream water cushion therefore has time to form before maximum overflow occurs.

The overflow levee is wide and its downstream face has a gentle slope with no riprap protection. It has been flooded several times without signs of erosion on the downstream face. Riprap can be added to the upstream part of the levee if necessary.

Example 2: The Printegarde plain in Livron-sur-Drôme

After the CNR dam in Baix-Le-Logis-Neuf was commissioned, the Printegarde plain was flooded in the least damaging conditions possible for the land, crops, plantations and even buildings in the exposed areas. We described its operation in Chapter 3 to illustrate the advantages of using a 2D model.

The work carried out in 1960 includes the following structures (Figure A3.3):
– An overflow-resistant levee beginning on the Drôme River in the south and closing transversely in the north (brown dotted line).
– A spillway on the levee at the upstream junction of the Petit-Rhône River, which begins to overflow when the Rhône's flow rate reaches 5,000 m³/s.
– A siphon in the spillway starts operating at a water level corresponding to a flow rate of 4,600 m³/s.
– A movable dam at the downstream end of the Petit-Rhône River to balance the water levels between the floodplain and the flooded Rhône; it is opened for Rhône River flows of around 6,200 m³/s.
– Downstream, a siphon under the Drôme River to facilitate the progressive flooding of the flood expansion area from downstream and to drain the plain when the water level goes back down. The siphon is closed when the river's flow reaches 5,500 m³/s.

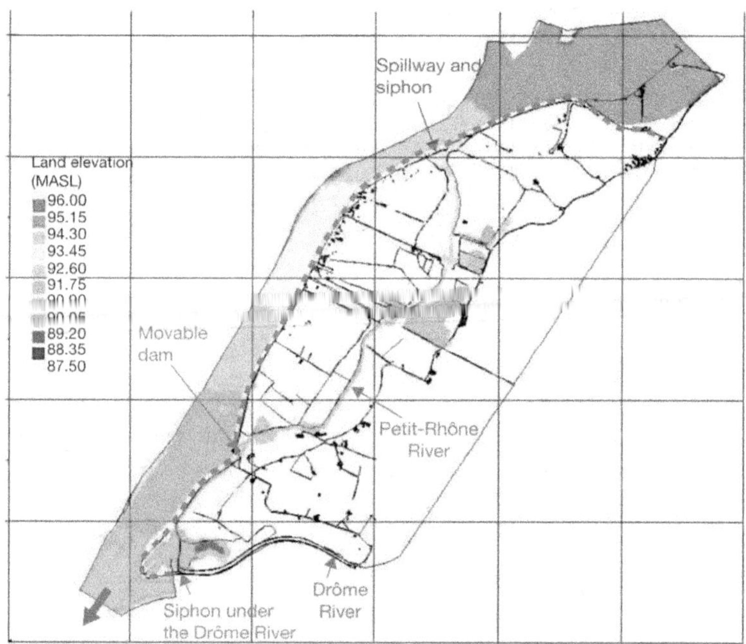

Figure A3.3. Beginning of upstream flooding of the Printegarde plain.

Glossary

Derived from G. Degoutte, with French translation in italics.

Levee *(digue):*

Raised, predominantly earthen, human-made structures designed to protect against flood events in (low-lying) coasts, rivers, lakes, and artificial waterways. Levees are not reshaped by natural actions such as current or wave and wind action (therefore exclude dunes).

Levees are also called dikes, dykes, embankments, flood banks, or stopbanks according to local usage.

1D models *(modèles 1D)*

In 1D (or one-dimensional) hydraulic models, flow is considered to be straight enough so that each section will be perpendicular to the flow axis and defined according to the abscissa.

2D models *(modèles 2D)*

2D (or two-dimensional) hydraulic models consider the water surface's cross slope in bends. They help measure the range of velocities (unlike 1D or storage cell models).

Alluvium *(alluvions)*

Alluvium refers to fine or coarse particles that are either deposited or carried away by the river's current. It covers the substratum made of hard or relatively soft rock (schist, sandstone, marl, etc.). A river typically flows over its alluvium.

Apron *(radier)*

An apron is the lower, mainly horizontal part of a water transport structure (canal, flood gate chute) or movable weir. It is typically made of concrete or masonry. The term is occasionally used to refer to the bottom of a watercourse with rapid flow.

Backward aggradation *(exhaussement régressif)*

When the river's longitudinal profile is raised in the upstream direction. It is typically caused by human actions that elevate the streambed in a specific location (such as weir construction). It may also have natural causes, such as sediment inflow coming from a tributary with supercritical flow during a flood.

Backward erosion *(érosion régressive)*

A mechanism causing the streambed to collapse in the upstream direction. It is typically caused by human actions that lower the streambed in a specific location (meander cut off, weir removal, extraction of alluvium from the riverbed, etc.).

Bank erosion *(érosion de berge)*

A mechanism in which soil particles in the riverbank are carried away. The most common occurrence is when the current erodes a bank by displacing and carrying away particles via the shear stress of the water flowing over the banks. Other types of bank erosion include trampling by cattle, damage caused by burrowing animals (coypus, badgers, crabs, etc.), or runoff erosion. Bank erosion caused by the current is just one mechanism of riverbed deformation; others include sliding and sloughing.

Batter *(fruit)*

Batter is the receding slope of a wall or structure.

Bed load transport *(charriage)*

Granular materials carried away by bed load transport will roll, slide, or hop along the bottom of the riverbed. Bed load particles move at a much slower pace than water, less than one metre per hour.

Berm *(risberme)*

Horizontal flat section built into a slope, such as the upstream or downstream slope of a levee.

Catchment area (or watershed) *(bassin versant)*

The catchment area (or watershed) of a watercourse supplies it with surface and ground water.

Caving (of a river bank or a levee toe) *(sapement)*

See "scour".

Chute *(coursier)*

A channel designed to transfer water from the spillway to the downstream toe of the levee (or dam).

Civil engineering *(génie civil)*

Civil engineering is the art of building. It mainly applies to buildings and public works (transport infrastructure, hydraulic works: dams, levees, canals, harbours, etc.).

Crest (of a hydraulic work) *(crête d'un ouvrage hydraulique)*

The crest of a hydraulic structure (dam, levee, weir, spillway) is its highest horizontal (or nearly horizontal) part. It is normal for the water level to reach the crest in the case of spillways or weirs, but this is dangerous in the case of dams or earthen levees.

Crisis *(crise)*

A crisis begins as soon as water flows into the protected area. When flooding occurs, stakes in the protected area are at risk. Public authorities are responsible for managing the crisis (mayor, prefect, State). (See "emergency")

Current-driven erosion *(érosion par le courant)*

A mechanism in which the materials in the river system channel walls are torn and carried away by the current. Current-driven erosion may occur on banks, the channel bottom, or both. See "bank erosion", "backward erosion", "progressive erosion".

Dam *(barrage)*

Dams are structures that block a catchment area and store a volume of water. They are different to river weirs designed to raise the waterline, which only partially block the riverbed. Dams are made of earth, riprap, masonry, or concrete. Earthen dams are also sometimes called levees, but this term is incorrect and should be avoided. A dam is built across a river to block at least the riverbed, and frequently the floodplain and beyond. A levee, on the other hand, never blocks the riverbed.

Danger flood *(crue de danger)*

A danger flood might cause the levee to break in at least one part of the area under consideration.

Diversion *(dérivation)*

Continuously or temporarily rerouting water from a river (or a canal or reservoir). Diversion will always be caused by gravity.

Dynamic flood retention *(ralentissement dynamique)*

The goal of dynamic flood retention is to mitigate flooding by restraining flows before they reach the watercourse. This process uses the absorption capabilities of floodplains and temporarily stores a portion of flood volumes in specific hydraulic structures. Actions or structures that contribute to dynamic flood retention include adaptations to the catchment area (cultural practices, hedges, etc.), the urban environment (reservoir structure carriageways, etc.), riverbeds or floodplains (increased roughness), or hydraulic works (basins, open sluice dams, etc.). (Chastan et al., 2004)

Emergency *(urgence)*

In this guide, an emergency corresponds to a situation in which a levee experiences hydraulic load due to a flood. During an emergency, the levee manager is responsible for operations. The state of emergency stands until the water level decreases or the waterline reaches the levee system's protection level. When the waterline is above the protection level (meaning that water is flowing into the protected area), this is called a crisis. See "crisis".

Erosion *(érosion)*

When soil or rock particles are torn away and transported by wind, waves, rain, snowmelt and freeze-thaw cycles, or via sliding ground, rockfall, avalanches, or debris flow. In a riverbed, erosion may be isolated (bank erosion in a bend or downstream of a weir) or generalised (backward or progressive erosion).

Face *(parement)*

The outer visible part of a structure (dam, weir, or levee).

Flashy tributary *(affluent réactif)*

A flashy tributary reacts quickly to any rain event. Rain flows almost directly to the river due to a lack of infiltration (such as soil sealing in urban areas) or a high average slope in the catchment area.

Flood *(crue)*

A flood is an elevation in the level of a watercourse resulting from the arrival of large volumes of water due to rainfall, snowmelt, or both. A flood can also have a less natural cause, such as dam or levee failure or gates opening at the wrong time. See "flooding". When both a flood and flooding affect the same area, the flood will be the cause and flooding will be the result. We traditionally distinguish between slow floods that affect large watercourses with extensive catchment areas and rapid floods caused by very heavy and isolated rainfall.

Flood attenuation *(écrêtement d'une crue)*

A flood is attenuated when its volume is temporarily stored in a reservoir that is either natural (lake, natural flood expansion area) or built (dam reservoir, controlled flood expansion area, basin). See the definition for "flood hydrograph flattening", which has a similar meaning and is sometimes seen as synonymous.

Flood expansion area *(zones d'expansion des crues)*

A flood expansion area is a natural or leveed area into which water spreads when a watercourse overflows into the floodplain. Temporary water storage attenuates the flood by extending its flow time. This storage helps aquatic and land ecosystems function properly. A flood expansion area typically refers to areas with no or limited urbanisation and development. It is also mentioned in two French circulars: 26 April 1994 on flood prevention and management in flood-prone areas and 24 April 1996 on provisions applicable to existing constructions and hydraulic structures in flood-prone areas.

Flooding *(inondation)*

Flooding occurs when an area is submerged by water. In continental environments, it is caused by high floods (natural floods due to rain, snowmelt, or a glacier outburst; or artificial floods due to the failure of a dam or levee or an ill-timed gate operation). It can also result from substantial runoff (mainly in cities). Not all flooding events are caused by overflowing watercourses. Flooding and floods are different concepts. See "flood".

Flood hydrograph flattening *(laminage de crue)*

This term is often used as a synonym for "flood attenuation", but there are differences between the two concepts. **Attenuation removes part of the flood volume:** a reservoir that is empty before the flood arrives either stores the entire flood volume or lowers the flood peak level.

Hydrograph flattening maintains the flood volume. A flood hydrograph is flattened when its volume is **temporarily** stored in a reservoir that is either natural (floodplain, lake) or artificial (dam reservoir, basin). A reservoir that is full before the flood arrives will flatten its hydrograph.

In short, attenuation lowers the flood peak while hydrograph flattening, as its name suggests, flattens it.

Floodplain *(lit majeur)*

A floodplain is a flood-prone area defined by the highest water level. This handbook uses the term floodplain to describe the area surrounding the riverbed, whereas other authors include the riverbed in the floodplain.

Freeboard *(revanche)*

The height difference between the crest of a work (dam, levee, canal) and the water surface or waterline for the situation in question.

Free flow *(écoulement dénoyé)*

Flow over a spillway or weir is considered "free" if the downstream water level does not affect the upstream level. A spillway always has free flow for low flow rates but may have submerged flow for high flow rates.

Fuse plug *(fusible)*

See "fuse plug spillway" in Appendix 1.

Gabion *(gabion)*

Gabions are structures made of wire mesh cages filled with rocks or pebbles. They typically have a parallelepiped shape. Gabions are used to build retaining walls along thoroughfares or in mountain environments or to build weirs or groynes. To protect river or canal banks, Reno mattresses (see definition) are preferable. Gabion is derived from the Italian word *gabbioni*, meaning "large cage", since the process was invented in Italy in the 16th century. Those gabions were large wicker baskets filled with earth to protect banks or strengthen military positions.

Geotextile *(géotextile)*

In civil engineering applications, a geotextile is a rot-proof, flat, and permeable textile made of (natural or synthetic) polymer. It is produced as unwoven, knitted, or woven flexible mats that are placed in direct contact with the ground or other materials. The most commonly used polymers are polypropylene, polyethene, and polyester. An unwoven geotextile comprises fibres, filaments, or other items placed in a specific pattern or at random and bound by a mechanical, thermal, or chemical process. A woven geotextile is produced by intertwining one or more threads, filaments, or other items.

Gravity levee *(digue poids)*

A gravity levee is a structure that will resist water pressure thanks to its weight (analogous to gravity dams). It is made of masonry or concrete.

Hazard *(aléa)*

A hazard is an undesirable phenomenon (or event) expressed as the relationship between intensity and frequency. This term typically refers to natural physical phenomena (flooding, avalanches, earthquakes, falling blocks, etc.). With flooding, a hazard is expressed as its probability of occurrence (or return period) at a given flood intensity. A hazard is one aspect of "risk" (the other is "vulnerability").

Hydraulic head *(charge hydraulique)*

Hydraulic head is energy per unit of liquid weight. The head at point P, in comparison to a reference horizontal plane, is $H_p = z_p + p / \gamma_w + V^2 / 2g$, where z_p is the point's elevation, γ_w is the liquid's unit weight, and p and V are the pressure and velocity measured at this point.

Hydraulic jump basin *(bassin de ressaut)*

See "stilling basin"

Hydrograph *(hydrogramme)*

The variation in flow over time shown as a curve (most often) or a series of numbers or functions.

Inflatable weir *(seuil gonflable)*

Inflatable weirs are movable devices that will help limit a weir's upstream impact during high water levels. An inflatable weir comprises a reinforced elastomer flexible membrane placed on a concrete apron. The weir is inflated either with water or air. When the upstream water level increases, the membrane will gradually subside. Re-inflation may be triggered manually or automatically.

Internal erosion *(érosion interne)*

A mechanism that causes the particles in an earthen hydraulic structure or its foundation to be dragged away under the influence of a hydraulic gradient.

Jam *(embâcle)*

A jam is a pile of trees or other floating objects or blocks of ice that have been carried away by the current and are blocked somewhere in the riverbed. A jam may block all or part of the riverbed. Jams usually occur in continuous sections of the river or occasionally upstream of bridges or weirs during high floods.

Labyrinth weir *(seuil labyrinthe)*

A labyrinth weir (or spillway) is a sequence of small duckbill spillways. The total length of this weir is typically three to five times its width or more. If the head reaches half the wall height, the flow per metre will be twice as high.

Leveed area *(val)*

Designation used for (large-scale) leveed systems, particularly on the Loire River. The levee is designed to protect the leveed area. There are 33 leveed areas on the Loire River.

Leveed system *(système endigué)*

A complete system comprising one or more levees and the area they are meant to protect.

Local Emergency Action Plan *(plan communal de sauvegarde)*

Municipalities with approved Risk Prevention Plans (PPR in French) must create a Local Emergency Action Plan (LEAP). This plan, as described in French Decree 2005-1156 of 13 September 2005, is incorporated into overall emergency planning.

Movable weir *(seuil mobile)*

See "variable spillway weir" in Appendix 1.

Overflow *(surverse)*

As is standard, we use the term "overflow" to refer to water flowing over a levee crest into an area not designed for that purpose. When water flows over a spillway, we use the term "discharge".

Overflow levee *(digue déversante)*

For some, this term refers to a levee spillway. For others, it refers to an overflow-resistant section of a levee. To avoid ambiguity, the handbook clearly distinguishes between the terms **overflow-resistant levee** and **levee spillway** where applicable.

Overflow-resistant levee *(digue résistant à la surverse)*

This kind of levee will not be damaged by the passage of water. It can be made of concrete or masonry. It can also be made of armoured earth, which means that it is covered with material that will resist the passage of water, such as embedded stone masonry or concrete riprap. In either case, the ground at the levee toe must resist jet erosion either naturally (due to rock composition), by extending the lining, or through adaptations.

Perched riverbed *(vallée en toit)*

A perched riverbed is a plain or valley in which the riverbank is the highest point. *See Figure G1.1*

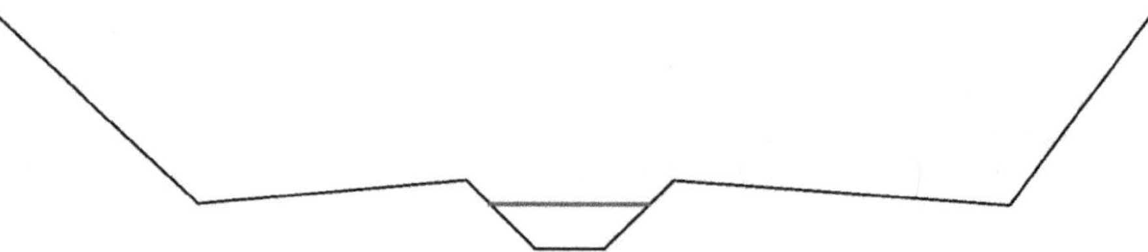

Figure G1.1. Cross-section of a perched valley or riverbed.

Piano key weir *(seuil en touches de piano)*

Piano key weirs (or PK weirs), designed by F. Lemperrière (Hydrocoop), increase a weir's conveyance for a given footprint. They are not patented. Unlike labyrinth weirs, the flow will include two nappes: a bottom jet flowing along the tilted apron of the downstream cavity, and an artificial nappe that helps ventilate the spillway through an overhang. Conveyance can be three times higher than for a straight weir with the same footprint.

PK weir

See "piano key weir".

Progressive aggradation *(exhaussement progressif)*

When the river's longitudinal profile is raised in the downstream direction. It is typically caused by human actions that lead to a surplus of alluvium materials in the riverbed (diversion via a canal, removal of a weir, etc.).

Progressive erosion *(érosion progressive)*

When the bottom of the watercourse collapse in the downstream direction. It is typically caused by human actions that lead to a shortage of alluvium materials (construction of a dam or weir, removal of alluvium from the riverbed, etc.).

Protected area (meaning protected from flooding) *(zone protégée)*

A protected area is a contiguous area within the floodplain that is sheltered from floods by a set of levees or other works (road embankments, etc.) or raised topographical features such as a hillside, promontory, or terrace. This area is considered flood-prone when there is no levee and protected up to a certain flood level when there is an intact levee. A leveed flood expansion area is also a protected area.

Protection flood *(crue de protection)*

The protection flood is the level at which water will start flowing over the spillway. The safety flood (see definition) may not be lower than the protection flood; otherwise, it is only an apparent protection flood. Without a spillway, the protection flood is typically considered the same as the safety flood as it cannot be specifically defined.

Reno mattress *(matelas Reno)*

A large-scale structure similar to a gabion (see definition) of limited thickness (20 cm to 30 cm). The name comes from the Reno River, a tributary of the Pô River.

Risk *(risque)*

Risk is the evaluation of a hazard in relation to the occurrence of an adverse event and its consequences. For natural risks, we consider both hazard and vulnerability. The concept of risk is not necessarily negative. For instance, a person camping next to a torrent accepts the risk of being flooded to enjoy the landscape. Structures built to protect against flooding are often accused of having a negative effect since they create confusion between risks and the resulting hazard. For example, building a levee on a river will reduce the flooding hazard, but if this increases the number of constructions, the vulnerability will increase and so may the risk. See "hazard" and "vulnerability".

Risk assessment *(étude de dangers)*

The levee manager is responsible for performing an assessment to define the risk levels that have been considered, describe the measures needed to mitigate these risks and identify the residual levels once these measures have been implemented. This assessment also considers all the risks associated with floods and the potential effects of a failure in the structures and any accidents or events relating to routine operation. Risk assessment is based on a risk analysis.

Risk Prevention Plan *(plan de prévention des risques)*

As established by the 2 February 1995 French law, a Risk Prevention Plan (RPP) considers all predictable natural hazards, flooding, avalanches, ground movements, forest fires, earthquakes, etc. The RPP falls under the government's responsibility. It should include prohibitions and requirements for exposed areas based on hazard type and intensity, as well as for indirectly exposed areas where the construction of new structures might increase or create new hazards. Following a public inquiry, input from the municipal councils concerned, and approval by the prefecture, the RPP becomes a public easement that allows for the application of criminal penalties.

Riverbed *(lit mineur)*

The area in which standard floods will flow, separated from the floodplain by the banks. For rivers with multiple branches (such as braided rivers), the riverbed will include the various channels and shoals or islands separating or running along the branches.

River geomorphology *(géomorphologie fluviale)*

The science that studies the shape that watercourses may take due to the flow of water and how they are created and change over time.

River weir *(seuil en rivière)*

Masonry, concrete, gabion, riprap, or wooden structures designed to raise the waterline, A river weir may use gravity to orient flow from a diversion canal, use water power, or have pumps. It may also create a small water body for leisure purposes. Bottom weirs can be built to prevent backward erosion.

Safety flood *(crue de sûreté)*

A safety flood is the level at which the entire levee will maintain a margin of safety in response to various means of failure (freeboard for waves and a sufficient safety coefficient to ensure stability). Once the safety flood is exceeded, the levee may fail.

Scour *(affouillement)*

Scour refers to the erosion caused by the current at the base of a riverbank, bridge pier, or other structure built in the centre or along the riverbed.

Sediment *(sédiments)*

Solid deposits carried away by water.

Sheet piles *(palplanches)*

Sheet piles are metal beams that are driven into the ground one by one. Each sheet pile is attached to the previous pile across its entire length thanks to the beam shape. Sheet piles used to be made of wood.

Sidewall *(bajoyer)*

A sidewall is the generally vertical part of a structure such as a weir, spillway, canal, or lock that separates the structure from the bank or embankment and serves as a retaining wall.

Slope *(pente)*

The slope of a bank or levee face is the ratio of the difference in altitude between two points and their horizontal distance. It is therefore a vertical measurement in relation to a horizontal measurement. The opposite is often claimed, in error. This can lead to severe distortions, such as claiming that a 3:1 slope is lower than a 2:1 slope. Adding the letters H and V cannot correct this mistake. An earthen levee typically has a slope of 1:2 (or 1V:2H) but never 2:1 (or 2H:1V). The opposite of slope is batter.

It is preferable to use the term slope to refer to horizontal measurements and batter for vertical measurements.

Spillway *(déversoir)*

A spillway is an addition to a hydraulic structure that water will flow over in normal circumstances or just during floods. A spillway may be a flood spillway designed to protect a dam or a levee. For dams, it will help reduce the frequency of overflows on other parts of the work. For levees, it may also help mitigate the effects of these overflows. When a spillway is on a river, canal, or ditch, it becomes possible to measure the flow rate.

Stakes *(enjeux)*

Stakes refer to individuals, properties, businesses, resources, heritage, etc. that might be affected by a (natural or anthropogenic) event. See also "risk", "hazard", "vulnerability".

Stilling basin *(bassin de dissipation d'énergie)*

A stilling basin is created downstream of a river weir (or a dam or levee spillway chute, or a gate outlet). It dissipates water energy at the location of the hydraulic jump downstream of the junction between supercritical flow and subcritical flow.

Storage cell *(casier)*

A storage cell is an area within the floodplain that is surrounded by levees and potentially by other raised embankments and natural relief elements. It is a homogenous area in terms of overflow, in which water is stored but does not flow (unless it is filled with water, then drained). It does not contribute to flows, unlike floodplains with barriers across them. This term should not be used when referring to an area's land use and purpose. Instead, the terms leveed flood expansion area or protected area should be used. The term storage cell only refers to hydraulics: a hydraulic storage area where water is practically horizontal and which connects to other storage cells or the watercourse via flows such as water discharge, openings, or siphons. This is typically the meaning used in storage cell models (see definition).

Storage cell models (or 1D+ models) *(modèles à casiers)*

As an alternative to 1D models, storage cell models consider floodplain areas, called storage cells, whose contours are based on topography (hillsides, levees). These models require a consistent water level across the storage cell and are designed as 1D models. They are sometimes incorrectly called 1D+ models, which could be misleading.

Subcritical flow *(écoulement fluvial)*

A channel has subcritical flow when the Froude number is below 1. Subcritical flow is controlled from downstream.

Streaming flow *(écoulement noyé)*

A spillway or weir has streaming flow if the downstream water level affects the upstream level. In this case, there is no more supercritical flow. A spillway may operate with a streaming flow for high flow rates, but never for low flow rates.

Supercritical flow *(écoulement torrentiel)*

A channel will have supercritical flow when the Froude number is above 1. Supercritical flow is controlled from upstream.

Unprotected floodplain *(franc-bord)*

Section of the floodplain between the watercourse and a levee.

Vulnerability *(vulnérabilité)*

The vulnerability of an environment, property, or individual is the probability that it will sustain damage resulting from a natural or anthropogenic event. It is an assessment of all the predictable effects of such an event on these "stakes" (see definition). For instance, inhabited areas are more vulnerable to flooding hazard than farming areas. After a flooding hazard (flood with a specific return period), the vulnerability of an urban area is determined by the cost of the damage (in euros). Vulnerability is one component of risk; the other is hazard.

See "risk" and "hazard". To reduce vulnerability, we can attempt to reduce the negative impact of "passive" vulnerability factors (improving the resistance of constructions, reducing exposure by building protective structures, etc.). We can also adjust "active" vulnerability factors through stakeholder actions and by improving management practices.

Wash load *(auto-suspension)*

The transport of catchment area materials that are smaller than fine particles in the riverbed. This process is not related to the bed transport capacity. It is called wash load as opposed to "bed load", the transport of particles along the bed.

Reference List

Barthes, H. et al. (1994) Tunnels – the Fond-Pignon discharge site. In: *Proceedings of the Institution of Civil Engineers – Civil Engineering, Channel Tunnel, Part 3: French Section.* London, Thomas Telford Ltd, pp. 23–25.

Benhamed, N., Chevalier, C., Bonelli, S. (2012) Chapitre 5: Érosion par écoulement localisé dans un conduit. In: *Érosion des géomatériaux*, coll. Traité MIM, Hermès Science Publication/Lavoisier.

Bravard, J.P., Clemens, A. (eds.) (2008) *Le Rhône en cent questions.* Villeurbanne, *Édition* Graie.

CEMAGREF/CETMEF (2007) *Guide méthodologique pour le pilotage des études hydrauliques*, Ministère de l'Écologie, du Développement et de l'Aménagement Durables, Direction Générale de l'Urbanisme, de l'Habitat et de la Construction.

CEPRI (2008) Évaluation *de la pertinence des mesures de gestion du risque inondation: Manuel des pratiques existantes.* Available on www.cepri.net.

Chanson, H. (1995) *Hydraulic design of stepped cascades, channels, weirs and spillways.* Oxford, Pergamon.

Chastan, B. (coordinator) (2004) *Le ralentissement dynamique pour la prévention des inondations* : guide des aménagements associant l'épandage des crues dans le lit majeur et leur écrêtement dans de petits ouvrages, Ministère de l'Écologie et du Développement Durable, CEMAGREF.

Cœur, D. (2008) Étude *historique: un* éclairage *sur l'histoire de l'aménagement de la basse vallée du Vidourle,* Synthèse, Syndicat intercommunal d'aménagement du Vidourle.

Cole, B., Ratanasen, C. Maiklad, P., Liggins, T. and Chirapuntu, S. (1977) Dispersive clay in irrigation dams in Thailand. From the British Library's *The world's knowledge.*

Dabling, M.R., and Tullis, B.P. (2012) Piano key weir submergence in channel applications. *Journal of Hydraulic Engineering* 138 (7), 661–666. Available from: doi:10.1061/(ASCE) HY.1943-7900.0000563

Degoutte, G. (2012) *Diagnostic, aménagement et gestion des rivières: Hydraulique et dynamique fluviales appliquées. 2nd edition.* Cachan, Tec & Doc – Lavoisier.

De Person, F. (2001) *La Loire à Blois, le déchargeoir de la Bouillie en action, ou le contrôle du chemin de l'eau de 1584 à 1907.* Édition DIREN Centre.

Dion, R. (1934) *Le Val de Loire. Étude de géographie régionale.* Tours, Arrault et Cie.

Doussin, N. (2009) *Mise en œuvre d'une stratégie globale de prévention du risque d'inondation: le cas de la Loire moyenne.* PhD thesis. Université de Cergy-Pontoise.

Erpicum, S., Laugier, F., Boillat, J. L., Pirotton, M., Reverchon, B. and Schleiss, A. J. (eds) (2011) *Labyrinth and piano key weirs – PKW 2011.* Leiden, CRC Press.

Forbes, P. J., *Sheerman-Chase, A., Birell, J.* (1980) Control of dispersion on the Mnjoli dam. In: *Water Power and Dam Construction. 32 (12), pp. 23–27.*

Fournier, M. (2005) *L'élaboration et la mise en* œuvre *du Programme Comoy (1867) sur la Loire moyenne: analyse du jeu d'acteurs et principaux enseignements.* PhD thesis. Université de Tours.

Goutx, D., Mériaux, P., Tourment, R. (2004) Conception hydraulique des déversoirs des endiguements de protection contre les inondations. In: *Colloque technique CFGB MEDD, Sécurité des digues fluviales et de navigation, 25-26 November 2004, Orléans.* CEMAGREF éditions, pp. 531-550. Available from the CFBR website.

Gutschick, K.A. (1978) Lime stabilization under hydraulic conditions. In: *4th lime congress, Hershey, PA*, pp. 1–20.

Hager, W. (1987) Lateral outflow over side weirs. In: *Journal of Hydraulic Engineering. 113* (4), pp. 491–504.

Hanson, G.J., Robinson, K.M. and Cook, K.R. (2001) Prediction of headcut migration using a deterministic approach. In: *Transactions of the ASAE.* 44(3), pp. 525–531.

Herrier G., Lesueur D., Pujatti D., Auriol J.-C., Chevalier C., Haguigui I., Cuisinier O., Bonelli S. and Fry J.J. (2012) Lime-treated materials for embankment and hardfill dam. In: *24th International Symposium on Dams for a Changing World, 6–8 June, Kyoto.*

Hinchliff, D. and Houston, K. (1984) *Hydraulic design and application of labyrinth spillways.* US Bureau of Reclamation. Denver, Colorado.

Ho Ta Khan, M., Sy Quat, D. and Xuan Thuy, D. (2011) PK weirs under design and construction in Vietnam. In: Erpicum, S., Laugier, F., Boillat, J. L., Pirotton, M., Reverchon, B. and Schleiss, A. J. (eds) *Labyrinth and piano key weirs – PKW 2011.* London, CRC Press, pp. 225–232.

Ho Ta Khanh, M., Chi Hien, T. and Thanh Hai, N. (2011) Main results of the PK Weir model tests in Vietnam (2004 to 2010) In: Erpicum, S., Laugier, F., Boillat, J. L., Pirotton, M., Reverchon, B. and Schleiss, A. J. (eds) *Labyrinth and Piano Key Weirs – PKW 2011.* London, CRC Press, pp. 191–198.

Howard, A.K., Bara, J.P. (1976) Lime stabilization on Friant-Kern Canal. In*: US Bureau of Reclamation Report No REC-ERC-76-20US.* Denver, Colorado.

Hughes, A. K. and Hoskins, C. G. (1994) A practical appraisal of the overtopping of embankment dams. In: *Reservoir Safety and environment, proceedings of the 8th conference of the British Dam Society held at the University of Exeter, 14–17 September 1994 London Thomas Telford, pp. 260–270.*

Hunzinger, L. (2007) The 2005 floods in Switzerland: lessons learnt for flood risk management. In: *Workshop on Community preparedness and public participation, 29–30 October 2009, Krakow, Poland.*

ICAT (2006) *Avis sur le plan Vidourle,* Ministère de l'Écologie et du Développement Durable, Direction de la Prévention des Pollutions et des Risques.

IRSN (Eyrolle, F., Duffa, C., Leprieur, F., Rolland, B., Antonelli, C., Marquet, J., Salaun, G. and Renaud, P.) (2004) *Conséquences radiologiques des inondations de décembre 2003 en Petite Camargue au lieu dit Claire Farine.*

Knodel, P. (1987) Lime in canal and dam stabilization. In: *US Bureau of Reclamation Report No. GR-87-10 Engineering and Research.* Denver, Colorado.

Laborie, V. and Ladreyt, S. (2005) *Notice sur les déversoirs – synthèse des lois d'écoulement au droit des déversoirs.* CETMEF.

Laugier, F. (2007) Design and construction of the first Piano Key Weir (PKW) spillway at the Goulours dam. *International Journal of Hydropower and Dams.* 14 (5), pp. 94–101.

Laugier, F. and Vermeulen, J. (2018) Quinze ans et onze pkw plus tard – retour d'expérience sur la conception et la construction d'évacuateurs de crue labyrinthes de type pkw. In: *Colloque CFBR: Méthodes et techniques innovantes dans la maintenance et la réhabilitation des barrages et des digues, 27-28 Novembre 2018, Chambéry.*

Lempériere, F. (2017) Low submersible dams. In: *International symposium of the 84th ICOLD Annual Meeting, 18 May 2016, Johannesburg.*

Lemperrière, F. and Ouamane, A. (2003) The piano keys weir: a new cost-effective solution for spillways. In: *Hydropower & Dams.* 7(5), pp. 144–149.

Lencastre, A. (1996) *Hydraulique générale.* Paris, Eyrolles.

Londe, P. and Lino, M. (1992) The faced symmetrical hardfill dam: a new concept for RCC. In: *International Water Power & Dam Construction.* 44 (2), pp. 19–24.

Mallet, T., Royet, P. and Cault, J.B. (2004) Reconstruction de la digue d'Aramon après la crue de septembre 2002. In: *Colloque Technique CFGB et MEDD, Sécurité des digues fluviales et de navigation.* Orléans, CEMAGREF éditions, pp. 429–445. Available on the CFBR website.

Maurin, J. and Pasquet, F. (2004) Les digues de Loire: utilisation du modèle hydraulique Loire moyenne pour le dimensionnement fonctionnel des levées et des ouvrages annexes. In: *Colloque Technique CFGB et MEDD, Sécurité des digues fluviales et de navigation.* Orléans, CEMAGREF éditions, pp.227-238. Available on the CFBR website.

Maurin, J., Ferreira, P., Tourment, R. and Boulay, A. (2012) Niveau de sûreté des digues: un outil pour l'évacuation massive du val d'Orléans en cas de crue majeure de la Loire. In: *Congrès SHF événements extrêmes fluviaux et maritimes, 1-2 February 2012, Paris.*

MEDAD (2007) *Évaluations socio-économiques des instruments de prévention des inondations, réalisées par le ministère en charge de l'Écologie sur la base d'une étude réalisée par Ledoux Consultants, le CEREVE, et le CEMAGREF.* Available on the CEPRI website.

Nerincx, N., Bonelli, S., Nicaise, S. and Herrier, G. (2019) The DigueELITE project: lessons learned and impact on the design of levees with lime-treated soils. In: *Hydropower & Dams.* (6).

Peyras, L., Royet, P. and Degoutte, G. (1991) Écoulement et dissipation sur les déversoirs en gradins de gabions. In: *La Houille blanche.* No 1–19910, pp. 37–47.

Pfister, M., Tullis, B. and Schleiss, A. (2013) Effect of driftwood on hydraulic head of Piano Key weirs. In: Erpicum, S. et al. (ed) *Proceedings of the 2nd International Workshop on Labyrinth and Piano Key Weirs II – PKW 2013, 20–22 November 2013, Paris.* Boca Raton, CRC Press, pp. 255–265.

Phillips, J. T. (1977) Case Histories of Repairs and Designs for Dams Built with Dispersive Clays. In: Sherard, J. L. and Decker, R. S. (eds). *Dispersive Clays, Relative Piping, and Erosion in Geotechnical Projects: a symposium presented at the 79th annual meeting, American Society for Testing and Materials, 27 June-2 July 1976, Chicago. pp. 330–340.*

Picon, B., Allard, P., Claeys-Mekdade, C. and Killian, S. (2006) *Gestion du risque inondation et changement social dans le delta du Rhône: les "catastrophes" de 1856 et 1993-1994.* CEMAGREF, Éditions QUAE.

Poligot-Pitsch, S. and Moreira, S. (2007) The French Experiment of an Inflatable Weir with Steel Gates. In: *Mitteilungsblatt der Bundesanstalt für Wasserbau 91.* Karlsruhe, Bundesanstalt für Wasserbau, pp. 93–98.

Pujol-Rius, A., Kollgaard, E., Brewer, G. and Eakin, J. (1991) Rehabilitation of Milner Dam. In: *17th International Congress on Large Dams, 17–21 June 1991, Vienna.* Q67/R23, pp. 373–397.

Rosier, B. (2007) *Interaction of a side weir overflow with bedload transport and bed morphology in a channel.* PhD thesis. École polytechnique fédérale de Lausanne.

Royet, P. and Mériaux, P. (2004) Les déversoirs fusibles le sont-ils vraiment? *Colloque technique CFGB MEDD, Sécurité des digues fluviales et de navigation, 25-26 November 2004, Orléans,* pp.187-199. Available on cfbr.eu.

Royet, P., Degoutte, G., Peyras, L., Lavabre, J. and Lemperrière, F. (2010) Cote et crues de protection, de sûreté et de danger de rupture. In: *La Houille blanche.* No. 2–2010, pp. 51–57.

Simons, D. B., Chen, Y. H. and Swenson, L. J. (1983) *Hydraulic test to develop design criteria for the use of Reno mattresses.* Fort Collins, Colorado, USA.

Sinniger, R. and Hager, W. (1989). *Constructions hydrauliques:* écoulements *stationnaires.* Lausanne, Presses polytechniques romandes.

SOGREAH (1962) Étude *systématique de déversoirs en* béton et de digues *déversantes revêtus d'un perré au mastic bitumineux.*

Tagwi, D. (2015) *Inflatable Weir Hydraulics.* MEng thesis. Stellenbosch University.

Cover photos
Left, Jargeau spillway on the left bank of the Loire (P. Royet)
Middle, Lunel spillway on the Vidourle River (G. Degoutte)
Right, masonry spillway on the right bank of the Giessen River (DDT Bas-Rhin)

Layout: EliLoCom

Imprimé pour vous par Books on Demand (Allemagne)